Bahmann Helge

A Network-Transparent, Retained-Mode Multimedia
Framework for Linux

Bahmann Helge

# A Network-Transparent, Retained-Mode Multimedia Framework for Linux

A Network-Transparent, Retained-Mode Multimedia Processing Framework for the Linux Operating System Environment

Südwestdeutscher Verlag für Hochschulschriften

**Impressum/Imprint (nur für Deutschland/ only for Germany)**
Bibliografische Information der Deutschen Nationalbibliothek: Die Deutsche Nationalbibliothek
verzeichnet diese Publikation in der Deutschen Nationalbibliografie; detaillierte bibliografische
Daten sind im Internet über http://dnb.d-nb.de abrufbar.
Alle in diesem Buch genannten Marken und Produktnamen unterliegen warenzeichen-, marken-
oder patentrechtlichem Schutz bzw. sind Warenzeichen oder eingetragene Warenzeichen der
jeweiligen Inhaber. Die Wiedergabe von Marken, Produktnamen, Gebrauchsnamen,
Handelsnamen, Warenbezeichnungen u.s.w. in diesem Werk berechtigt auch ohne besondere
Kennzeichnung nicht zu der Annahme, dass solche Namen im Sinne der Warenzeichen- und
Markenschutzgesetzgebung als frei zu betrachten wären und daher von jedermann benutzt
werden dürften.

Verlag: Südwestdeutscher Verlag für Hochschulschriften Aktiengesellschaft & Co. KG
Dudweiler Landstr. 99, 66123 Saarbrücken, Deutschland
Telefon +49 681 37 20 271-1, Telefax +49 681 37 20 271-0, Email: info@svh-verlag.de
Zugl.: Freiberg, TU, Dissertation, 2009

Herstellung in Deutschland:
Schaltungsdienst Lange o.H.G., Berlin
Books on Demand GmbH, Norderstedt
Reha GmbH, Saarbrücken
Amazon Distribution GmbH, Leipzig
**ISBN: 978-3-8381-0878-0**


**Imprint (only for USA, GB)**
Bibliographic information published by the Deutsche Nationalbibliothek: The Deutsche
Nationalbibliothek lists this publication in the Deutsche Nationalbibliografie; detailed
bibliographic data are available in the Internet at http://dnb.d-nb.de.
Any brand names and product names mentioned in this book are subject to trademark, brand or
patent protection and are trademarks or registered trademarks of their respective holders. The
use of brand names, product names, common names, trade names, product descriptions etc.
even without a particular marking in this works is in no way to be construed to mean that such
names may be regarded as unrestricted in respect of trademark and brand protection legislation
and could thus be used by anyone.


Publisher:
Südwestdeutscher Verlag für Hochschulschriften Aktiengesellschaft & Co. KG
Dudweiler Landstr. 99, 66123 Saarbrücken, Germany
Phone +49 681 37 20 271-1, Fax +49 681 37 20 271-0, Email: info@svh-verlag.de

Copyright © 2009 by the author and Südwestdeutscher Verlag für Hochschulschriften
Aktiengesellschaft & Co. KG and licensors
All rights reserved. Saarbrücken 2009

Printed in the U.S.A.
Printed in the U.K. by (see last page)
**ISBN: 978-3-8381-0878-0**

# Contents

| | | |
|---|---|---|
| **0** | **Introduction** | **5** |
| | 0.1 Typographic and diagram conventions . . . . . . . . . . . . . . . . | 7 |
| **1** | **Multimedia representation and processing** | **9** |
| | 1.1 Definition . . . . . . . . . . . . . . . . . . . . . . . . . . . . . . . . . | 9 |
| |     1.1.1 Audible media . . . . . . . . . . . . . . . . . . . . . . . . . . | 9 |
| |     1.1.2 Visual media . . . . . . . . . . . . . . . . . . . . . . . . . . . | 13 |
| | 1.2 Compositing and processing . . . . . . . . . . . . . . . . . . . . . . | 17 |
| |     1.2.1 Audio processing . . . . . . . . . . . . . . . . . . . . . . . . | 17 |
| |     1.2.2 Still image processing . . . . . . . . . . . . . . . . . . . . . | 18 |
| | 1.3 Representation in digital systems . . . . . . . . . . . . . . . . . . . | 20 |
| |     1.3.1 Techniques . . . . . . . . . . . . . . . . . . . . . . . . . . . . | 21 |
| |     1.3.2 Audio representation . . . . . . . . . . . . . . . . . . . . . . | 23 |
| |     1.3.3 Color representation . . . . . . . . . . . . . . . . . . . . . . | 27 |
| |     1.3.4 Image representation . . . . . . . . . . . . . . . . . . . . . . | 31 |
| |     1.3.5 Video representation . . . . . . . . . . . . . . . . . . . . . . | 34 |
| | 1.4 Compressed representations . . . . . . . . . . . . . . . . . . . . . . | 34 |
| |     1.4.1 Audio compression . . . . . . . . . . . . . . . . . . . . . . . | 35 |
| |     1.4.2 Image Compression . . . . . . . . . . . . . . . . . . . . . . . | 43 |
| |     1.4.3 Compressed video . . . . . . . . . . . . . . . . . . . . . . . | 45 |
| **2** | **Related work** | **53** |
| | 2.1 Media processing frameworks . . . . . . . . . . . . . . . . . . . . . | 53 |
| |     2.1.1 QuickTime . . . . . . . . . . . . . . . . . . . . . . . . . . . . | 55 |
| |     2.1.2 DirectShow . . . . . . . . . . . . . . . . . . . . . . . . . . . . | 63 |
| |     2.1.3 Network Integrated Multimedia Middleware . . . . . . . . . | 69 |
| | 2.2 Media processing in the Linux environment . . . . . . . . . . . . . | 72 |
| |     2.2.1 Low-level data capture and playback . . . . . . . . . . . . . | 73 |
| |     2.2.2 Media processing tools . . . . . . . . . . . . . . . . . . . . . | 77 |

# 3 Media processing framework architecture — 81
## 3.1 Design choices — 81
### 3.1.1 Processing model — 82
### 3.1.2 Data model — 83
### 3.1.3 Execution model — 83
### 3.1.4 Format transformations — 84
### 3.1.5 Component and object model — 84
## 3.2 Core architecture — 86
### 3.2.1 Modularization and component model — 87
### 3.2.2 I/O model — 88
### 3.2.3 Time model — 90
## 3.3 Media type support — 100
### 3.3.1 Audio — 101
### 3.3.2 Still images and video — 106
### 3.3.3 Compressed media — 109
### 3.3.4 User-defined representation types — 112
## 3.4 Processing — 112
### 3.4.1 Compositing — 113
### 3.4.2 Capture — 113
### 3.4.3 Rendering concept — 114
## 3.5 Documents — 119
### 3.5.1 Accessors — 120
### 3.5.2 Container file formats — 121

# 4 Cooperation with the X Window System — 123
## 4.1 Media processing extensions — 124
### 4.1.1 Timing and synchronization services — 128
### 4.1.2 Audio services — 130
### 4.1.3 Compressed media services — 139
## 4.2 Media presentation in the X Window System — 141
### 4.2.1 Video presentation — 141
### 4.2.2 Audio presentation — 143
### 4.2.3 Synchronization — 144
## 4.3 Renderer driver architecture — 145
### 4.3.1 General media rendering and synchronization — 146
### 4.3.2 Resource caching — 148
### 4.3.3 Handling of media elements — 149

## 5 System integration — 151
- 5.1 Bindings to audio programming interfaces .............. 151
  - 5.1.1 ALSA ................................. 151
- 5.2 Desktop audio mixing ........................... 152
- 5.3 GUI toolkit cooperation ......................... 154
  - 5.3.1 Media framework provisions ................. 155
  - 5.3.2 Gtk+/Qt bridge libraries .................... 157
- 5.4 Cooperation with other media frameworks ............. 158

## 6 Assessment — 161
- 6.1 Architecture model and API assessment ............... 163
  - 6.1.1 API field testing ......................... 164
  - 6.1.2 Comparison to QuickTime ................... 168
  - 6.1.3 Comparison to DirectShow ................... 173
  - 6.1.4 Limitations ............................ 179
- 6.2 Efficiency evaluation ............................ 181
  - 6.2.1 Overhead .............................. 181
  - 6.2.2 Audio latency and the X Window System .......... 187
- 6.3 Future work .................................. 189
  - 6.3.1 Future development of the "renderer" concept ...... 190
  - 6.3.2 X server infrastructure ..................... 191
- 6.4 Conclusions .................................. 192

## A Implementation notes: Media processing library — 195
- A.1 Data model .................................. 196
- A.2 Stream Demultiplexing .......................... 197
- A.3 BufferWindow concept ........................... 198
- A.4 Dynamic symbol lookup .......................... 198
- A.5 Pixel format and color space conversion ............... 199

## B Implementation notes: X Window System Extensions — 201
- B.1 Real-time audio processing ....................... 201
- B.2 Lock-free sample buffers ......................... 201

CONTENTS

# Chapter 0
# Introduction

This project started in 2002 with the goal of providing a comprehensive multimedia framework for the Linux operating environment. While it is not the sole project in this field, it is unique in the sense that it is the only one where clean and useful integration with the network transparent X Window System commonly used in this operating environment was an explicit goal from the very beginning. Most other framework designers have so far completely ignored the issue of network transparency, as it involves quite a number of technical challenges. The author explicitly wishes to demonstrate both the feasibility and the usefulness of network transparency in multimedia frameworks.

As it turned out during the evolution of the system design, this single requirement alone had a tremendous influence on the overall architecture. Ultimately the process of generalizing the key "use-cases" into design concepts led to an architecture that offers a sufficiently abstract data and processing model to support this scenario, while providing detailed control to the application – with remarkabale benefits even in scenarios that are unrelated to network transparency.

As part of this project an implementation of the multimedia framework and a considerable amount of supporting infrastructure was created, totalling about 80000 lines of code. While the implementation has been indispensable for continually trying out and evaluating different designs, it has over the course of this project left behind the seedling stage of a mere research prototype and blossomed into a usable piece of software. It has been used as the basis for several bachelor's theses as well as other projects, and has also been showcased publically.

While considerable research effort has been undertaken in recent years in the field of efficient multimedia data representation techniques (e.g. data compression and coding for storage or transmission), little has been published on the design of multimedia framework architectures in general. This situation is unfortunate, as core software design choices can become a limiting factor in terms of what capabilities can be added later (at least without "breaking" the design). It is an explicit goal of this work to close this "blind spot" by comparing and discussing design alternatives, and point out some weaknesses in many existing frameworks that generally either adopt an imperative immediate processing model, or a data-flow graph based approach (cf. chapter 2). These "traditional" design approaches strive to minimize the processing overhead imposed by the framework itself by keeping the abstraction level as low as possible, however they miss out on several possible optimizations.

The architecture that will be discussed is instead based on the ideas of *retained-mode* processing and *lazy evaluation*: "Retained" means that instead of executing processing operations, an internal abstract model representing the desired result is built up. "Lazy evaluation" means that the model is built up using functions and methods that "appear" to immediately affect the data in a way that programmers are used to from imperative or functional programming, but in reality only modify the retained representation. These concepts result in a considerably more complex design, but enable several high-level optimizations and provide a clean mechanism for delegating processing to provide network transparency. The experience gained with the provided implementation indicates that the overhead incurred through these concepts is so small as to be outweighed by far by the additional capabilities. The author therefore also wants to make the case for considerably more abstract processing and data models in multimedia frameworks.

This work is organized as follows: Chapter 1 gives a formal introduction of media representation and processing. It contains several precise mathematical definitions of how the underlying data is meant to be interpreted and transformed to make it clear that the media processing operations to be provided by a framework have a precise (if slightly idealized) model to which they should adhere and are not some ad-hoc collection of "best practice" methods. The author recognizes this is not the approach usually taken in the course of designing a multimedia framework, yet considers it essential as it provides an important yardstick to evaluate whether a desired process can usefully be mapped onto a target framework.

Chapter 2 will investigate the software architecture of existing media processing tools. This includes both comprehensive frameworks encompassing multiple aspects of media processing as well as singular libraries available in the target operating environment that are intended for specific narrow purposes (such as interfacing with playback and capture devices). This is not meant to be a rigorous discussion of any possible design but rather an overview of architectural ideas commonly found, highlighting some weaknesses from innocuous-looking design decisions to prepare the reader to better understand the "hows" and "whys" in the following chapter...

... number 3 which contains an architectural overview of the framework designed and implemented as part of this project. The chapter focuses on the "core" library of the media processing framework, providing the basic services and abstractions which the functional processing components plug into. It models the concepts discussed in chapter 1, but differs in important ways from the frameworks discussed in chapter 2. The reasoning behind the different architectural approach will also be discussed in detail here.

## 0.1. TYPOGRAPHIC AND DIAGRAM CONVENTIONS

The library introduced in chapter 3 is generic and agnostic with respect to types of media, sources for data capture, targets for playback, as well as formats for data representation. Instead it features a modular approach to provide these services through substitutable components. One particular such component, responsible for presentation through the X Window System, will be discussed in great detail in chapter 4. While this component does not have a "special" or more "intimate" relationship to the core library than any other component, it is instructive as it helps shed some light on the thought processes behind the "co-evolution" of the architectures for the core media processing library, the X interfacing component as well as several extensions to the X Window System also introduced in this chapter.

While the previous chapters discussed the media architecture mostly in isolation, chapter 5 takes a look from a different angle: How the architectural choices affect integration with the pre-existing software stack. This aspect is often overlooked unfortunately, leading to system designs that look good on paper, but are of limited usefulness because their design is too sealed off as to be combined with other important pre-existing frameworks in a single application.

Chapter 6 compares architectural features of the frameworks presented in chapters 2 and 3, including both qualitative as well as some quantitative results. The chapter will close with concluding remarks of what goals have been achieved and discuss possible directions for further research.

## 0.1 Typographic and diagram conventions

Mathematical formulas will always be set out in *italic* font, and function or variable names referenced within the text are set out in the same way. The following mathematical symbols will be used occasionally:

| Symbol | Meaning |
|---|---|
| sinc | Sinus cardinalis: $\operatorname{sinc}(x) = \lim_{x_0 \to x} \frac{\sin(\pi x_0)}{\pi x_0}$ |
| sgn | Sign: $\operatorname{sgn}(x) = x/|x|$ for $x \neq 0$, $\operatorname{sgn}(0) = 0$ |
| $\delta$ | Dirac's delta, defined by: $\int \delta(x) f(x) dx = f(0)$ |
| $x \mapsto f(x)$ | The function object performing the mapping: $f$ |

Most diagrams shown follow UML 2.0 notation. In the interest of brevity the types of diagrams, their elements and their semantics are not repeated here as they are extensively covered elsewhere (e.g. [44]). The reader is assumed to be familiar with this notation, as information contained in the diagrams is generally not repeated in the text. The diagrams are not even remotely intended to provide a complete and thorough specification of the objects or activities described therein as they will generally omit information that would be essential to complete an implementation from the model (e.g. private members, utility classes and functions) but that are of no relevance to the discussion.

While UML notation is generally preferred, sometimes short source code fragments are required for illustration. They will always appear in a paragraph or box on their own and will be set out in `Monospace` font. The text frequently references identifiers (used as class, variable or function names) found in source code examples or UML diagrams. To allow easy distinction of class, object or function identifiers from ordinary english text they are always set out in **SansSerif** font (following the convention used in UML as well).

# Chapter 1

# Multimedia representation and processing

The term *media* (and its singular *medium*) refers to content intended for reception by humans. This includes visual media (such as natural or synthetic still images, natural or synthetic video), audible media (natural or synthetic music, voices, noises etc.). It could also include content aimed at stimulating other human sensual organs (e.g. tactile or olfactory). However, due to lacking technical apperatus for recording, storage and presentation of these media types they are not in any widespread practical use, and the following discussion will therefore omit them. Media can also refer to completely abstract content that does not directly correspond to a perceptual organ, e.g. narrative, or text in general which can however be *delivered* through visual or aural organ. The term *digital media* (and its singular) refers to the digitized representation of *media*. The term *multimedia* refers to content integrating two or more different types of media.

## 1.1 Definition

The most prevalent types of media are audible media and visual media (still images, video). The following sections will formally introduce these media as well as mathematical models of media processing operations.

### 1.1.1 Audible media

The term *audible media* refers to content intended for reception by the human acoustic organ (the ear, see figure 1.1). The human ear is receptive to "rapid" changes in air pressure ("acoustic waves"): The acoustic wave entering the outer ear is received by the *tympanic membrane* and mechanically amplified through the *ossicles* in the middle ear. The amplified signal is then transferred to the fluid inside the *cochlea*. The cochlea is formed like a coiled tube, and it contains the *basilar membrane* which runs from the beginning to the end of the spiral

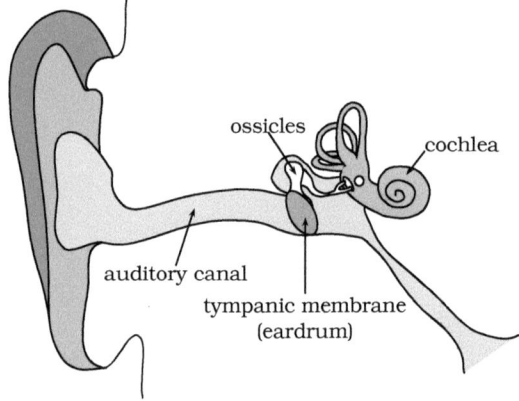

Figure 1.1: Schematic of the human ear

and acts as a "resonator": its width and stiffness varies along the length of the basilar membrane so that its resonant frequency varies as well. Hair cells placed on the basilar membrane detect motion and stimulate the auditory nerves[1] (see figure 1.1).

#### 1.1.1.1 Acoustics

The human ear can be understood as a mechanical real-time Fourier analyzer; it is sensitive in the frequency range of approximately 20Hz to 20kHz. Frequency discrimination is not uniform but follows a logarithmic scale: "Pure" sinusoidal oscillations (and hence sounds) are rare in nature, they are almost always accompanied by "higher" modes which are related to the base frequency through small integer numbers. On a logarithmic scale, higher modes have constant distance to the base mode thus this scale is better adapted to naturally occuring sounds.

An undamaged human ear is sensitive to changes in air pressure[2] somewhat

---

[1] Only the inner hair cells mainly act as "detectors" while the outer hair cells mainly act as "lock-in amplifiers" for oscillations. This mechanism is unique to mammals and helps to better localize oscillations on the membrane and thus provide better frequency resolution (but does not increase sensitivity).

[2] Pressure differences are indicated as "root mean square" (or "standard deviation"):

$$\bar{p} = \sqrt{\int p(t)^2 dt - \left(\int p(t) dt\right)^2}$$

somewhat below $2\,\mu\text{Pa}$ because energy of the wave is proportional to $\bar{p}^2$. For sinusoidal signals $\bar{p}$ is $1/\sqrt{2}$ times the maximum amplitude.

## 1.1. DEFINITION

below $2\,\mu\text{Pa}$ at the lower end – on the upper end of the scale it is "limited" only by the fact that high pressure differences will damage the ear (about $20\,\text{Pa}$ for short term exposure; long-term exposure at considerably lower intensities can already be harmful as well), up to the point were the eardrums may actually tear.

Air pressure is however not a good measure for "perceived loudness": Experimental results suggest that the logarithm of the energy of a wave provides a perceptually more uniform scale, thus "loudness" $L_p$ of sound is usually indicated in the logarithmic decibel scale:

$$L_p = 10 \log_{10}\left(\frac{p^2}{p_0^2}\right) \text{dB} = 20 \log_{10}\left(\frac{p}{p_0}\right) \text{dB}$$

where $p$ is the pressure of the sound to be measured and $p_0$ is the "zero" level (usually $2\mu\text{Pa}$ is chosen in acoustics, the unit is then written as $\text{dB(A)}$ to denote this).

Sensitivity is not uniform across the frequency range. Several empirical studies have determined "equal loudness contours" that measure threshold of perception and levels of perceived equal loudness across the frequency spectrum; the earliest work was performed by Fletcher and Munson [20], however ISO 226:2003 ("Acoustics – Normal equal-loudness-level contours") and most psycho-acoustic models used in audio compression are based on more recent results by Yôiti Suzuki et al [18].

Sensitivity may even change dynamically: The perception of sound may be affected by the presence of another sound – this effect is known as *auditory masking* in psycho-acoustics. Two main kinds of masking affect perception: Temporal masking, where previous sound reduces sensitivity for later sounds; and frequency masking, where sounds reduce sensitivity for concurrent sounds in adjacent frequency ranges.

The brain processes phase information only in a very limited fashion: Phase differences between signals of identical frequency in both ears are interpreted as if they were caused by different wave propagation latencies from a single audio source to the two ears. This assists in determining the "direction" of an audio source, it should however be noted that the brain also processes various other cues (such as amplitude differences, or low-pass filtering of waves travelling "around" the head) to localize an audio source.

#### 1.1.1.2 Audio signals

An audio signal will be defined as the pressure in a medium at a specified point in space:

**Definition 1** *A (continuous, monaural)* **audio signal** *A is a tuple* $(T_A, s_A)$ *where*

- $T_A = [t_{begin}; t_{end}]$ *or* $T_A = \mathbb{R}$ *is a temporal interval*

- $s_A : T_A \to \mathbb{R}$ is a function mapping each point in time $t$ to the pressure in a medium $s_A(t)$ at that point in time

Note that there is no requirement that the point of reference remains stationary. We will generally assume that the signal is "normalized" such that a value of zero corresponds to the "average" pressure (since the human ear is only sensitive to *pressure changes*).

### 1.1.1.3 Channels

A signal may represent the "input" signal for the human hearing apparatus (as if it had been recorded at the locus of the tympanic membrane), it may represent the "output" signal of a physical audio source (as if it had been recorded at the point of the source), or it may represent the signal at arbitrary other sampling points in space. The following definition incorporates this positional information:

**Definition 2** *An audio* **channel** *$C$ is a tuple $(s_C, p_C)$ consisting of*

- *an audio signal $s_C$*
- *a "location" $p_C$*

Note that the above definition uses the rather vague term "location" to describe the positional information included in the channel. This is intentional as different application scenarios have different requirements as to how the location is represented: Often purely logical designations (e.g. "left ear", "center", "somewhere to the right") are sufficient, while other scenarios require exact spatial coordinates. Thus $p_C$ serves as a "placeholder" that allows use-case-specific positional descriptions to be associated with an audio signal.

Reproduction of audio media is achieved by creating pressure waves at points in space corresponding to given audio signals (through loudspeakers): Listeners will perceive the superposition of these signals with their ears (see figure 1.2). In the absence of crosstalk between different audio signals (by proper spatial placement of speakers, e.g. headphone) two signals are sufficient to generate arbitrary stimulus. In general more positional audio sources are required to achieve a desired stereophonic effect.

The channel concept is used to represent both the "intent" of audio that is part of multimedia implementations as well as physically available capture and playback devices. Applications that want to reproduce audio may have to "map" the input signals to the output channels, i.e. they need to determine the contribution of each input signal to each output signal. This is easy when using headphones (which are crosstalk-free and the signals thus directly correspond to the input stimulus for the human ears), but in the general case this is not trivial.

## 1.1. DEFINITION

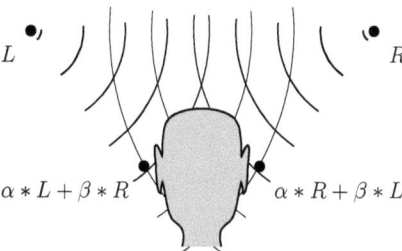

*Positional audio sources emitting two signals L and R result in the perception of $\alpha * L + \beta * R$ and $\alpha * R + \beta * L$ at the ears (where $\alpha$ and $\beta$ are transfer functions that describe both delay and attenuation). Note that a media processing application would have to perform the corresponding signal transformations if the position of speakers used to reproduce the audio does not match the positional intent of the signals L and R.*

Figure 1.2: Superposition of audio signals from two sources

### 1.1.2 Visual media

The term *visual media* refers to content intended for reception by the human visual organ (the eye). Physically the human eye is receptive to electromagnetic waves with wavelengths ranging from approximately 350 nm to 700 nm. The waves are received by different receptor cells in the retina: The *cone cells* responsible for daylight and color vision, and the (20-fold more numerous) monochromatic *rod cells* mostly responsible for night vision. Rod cells are extremely sensitive – in an undamaged human eye that has adapted to darkness a single photon per cell is sufficient to produce a perceptible excitation response.

#### 1.1.2.1 Color perception

Cone cells come in three flavors that exhibit different sensitivities across the spectrum of visible light (see figure 1.3); they are commonly referred to as L, M and S cone cells to designate that their peak sensitivity is in the long, medium or short wavelength range[3].

Human color perception has two important characteristics:

- Two spectra that result in the same excitation patterns of cone cells are indistinguishable (they appear to be the *same color* for humans).

---

[3] They are very often also referred to as 'red', 'green' and 'blue' cone cells which is considerably more memorable; however as figure 1.3 shows the absorbance spectra correspond only vaguely to these colors.

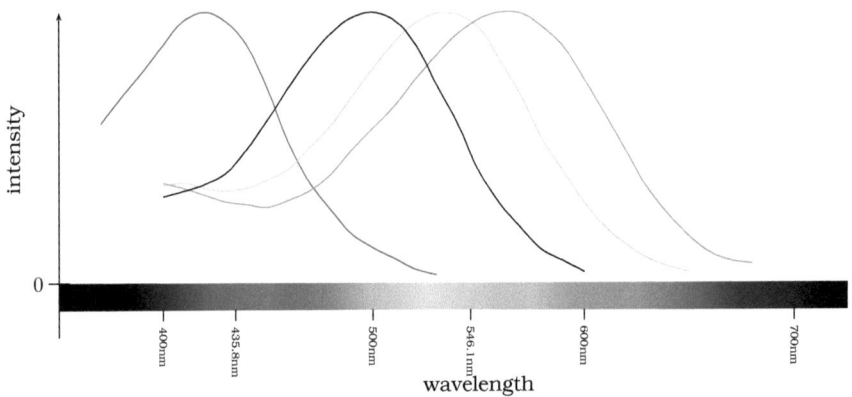

*The curves show the normalized relative excitation intensity of the following classes of cells: black – rod cells; blue – S cone cells ; green – M cone cells ; red – L cone cells, curves according to [9]. Note that the intensity peaks of the curves have been scaled to fit the diagram and do therefore not reflect absolute sensitivity.*

Figure 1.3: Absorbance spectra of different photoreceptor cells in the human eye.

- Superposition of two spectra results in the superposition of the excitation patterns[4] (this property is known as "Grassmann's law").

Both properties suggest the following mathematical model: Let functions $\overline{L}$, $\overline{M}$ and $\overline{S}$ correspond to the sensitivity curves of color cells as shown in figure 1.3. Let a spectrum be represented through a function $I$ mapping each wavelength $\lambda$ to its intensity $I(\lambda)$, then the triplet given by

$$(\langle \overline{L}, I \rangle, \langle \overline{M}, I \rangle, \langle \overline{S}, I \rangle)$$

(where $\langle X, I \rangle = \int_\lambda X(\lambda) I(\lambda) d\lambda$ denotes the inner product) describes the stimulus response of the three cell types, and we could denote the color impression of the spectrum $I$ using this triplet[5].

### CIE XYZ

We can interpret this process as a projection of the vector space of light spectra onto a three dimensional subspace[6]. Since the choice of basis vectors is however completely arbitrary we can use without loss of generality the following

---

[4] ignoring slight non-linearities due to saturation effects at high intensities

[5] Sensitivity maxima for cells of the same kind differ by up to $10\,\text{nm}$ [9] in the same individual, and even larger differences in the populace at large; moreover some humans possess only two kinds of cone cells and thus can distinguish less colors, and reports indicate to rare occurences of tetrachromats; thus the definition refers to an idealized model of an "average" human's color perception.

[6] defined through the vectors $L$, $M$ and $S$ of the dual vector space

## 1.1. DEFINITION

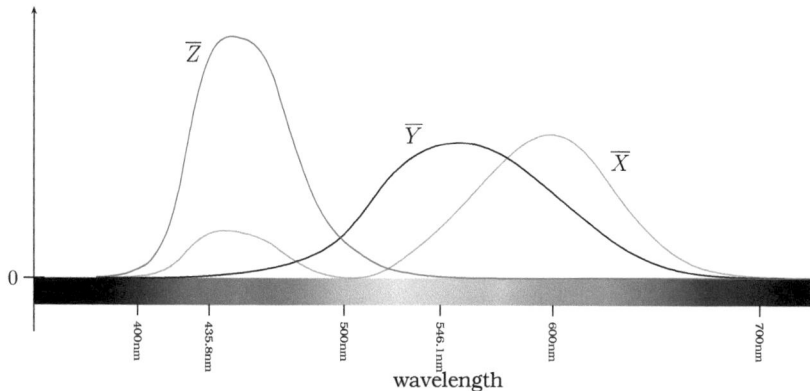

Figure 1.4: CIE XYZ color matching functions

**Definition 3** *(CIE XYZ) The* **color** *corresponding to the spectrum $I$ (describing the intensity $I(\lambda)$ of light with wavelength $\lambda$) is identified through the triplet of non-negative real values given by $(\langle \overline{X}, I \rangle, \langle \overline{Y}, I \rangle, \langle \overline{Z}, I \rangle)$; the functions $\overline{X}, \overline{Y}$ and $\overline{Z}$ ("CIE XYZ color matching functions") are given through figure 1.4 .*

This defines the CIE XYZ color model [12] based on research by Wright 1928 and Guild 1931[7]. In the following we will always write $(X, Y, Z)$ as a short-hand for a color triplet as per definition 3. The following properties should be noted:

- The triplet $(0, 0, 0)$ denotes the absence of light, i.e. "black".

- The definition as given is unique only up to a scaling factor that may be chosen arbitrarily. Uniformly scaling a triplet $(X, Y, Z)$ results in a color that perceptually is the same "tone", but differs in "intensity". Without loss of generality we can therefore assume that the scale is normalized such that $[0; 1]^3$ contains all colors we want to represent.

- For all physically reproducible colors, the triplet $(X, Y, Z)$ is non-negative (since $I(\lambda) \geq 0$ for every physically emittable spectrum $I$, and the color matching functions are non-negative as well). However, not every triplet $(X, Y, Z)$ corresponds to a physically reproducible color.

- The definition allows to decide when two spectra are perceived as the same colors, but not (directly) how to physically reproduce a color described by a given triplet

---

[7]Test subjects were asked to adjust the intensities of three monochromatic light sources at 700 nm, 546.1 nm and 435.8 nm wavelength until the superimposed result would match a given fourth monochromatic color; the collected test data experimentally established a relationship of light spectra that in turn allowed the construction of (triplets of) basis functions.

- The $Y$ component measures perceived "brightness" or *luminance* of a color.

The CIE XYZ color model as defined here is widely used in literature to objectively specify colors. Sometimes colors are specified through $(x, y, Y)$ instead with

$$x = \frac{X}{X+Y+Z} \quad , \quad y = \frac{Y}{X+Y+Z}$$

from which the original triplet $(X, Y, Z)$ can be derived; this representation separates chromaticity $(x, y)$ from luminance $Y$ and is referred to as CIE xyY. (Frequently only $(x, y)$ are given if luminance is of no importance or can be inferred in other ways.)

#### 1.1.2.2 Still images

Images will essentially be defined as "colored rectangular areas"[8]. However, images are often used as intermediate data for compositing operations, and thus we want to be able to associate additional information to every point of the image: This additional information consists of an "$\alpha$ value" that does not contribute to the chromaticity information but describes the compositing behavior. Intuitively it can be understood as "opacity" of the image at the specific point that affects the outcome if multiple images are "layered" on top of each other (see section 1.2.2 for a discussion how the $\alpha$ value is used in practical image compositing).

**Definition 4** *An **image** $I$ is a quadruple $(w_I, h_I, c_I, \alpha_I)$ where*

- *$w_I$ and $h_I$ are non-negative numbers*
- *$c_I$ is a mapping $[0, w_I) \times [0, h_I) \to [0, 1]^3$*
- *$\alpha_I$ is a mapping $[0, w_I) \times [0, h_I) \to [0, 1]$*

The definition should be interpreted as follows: An image assigns to every point $(x, y)$ of the rectangular area $[0, w_I] \times [0, h_I]$ a color $c_I(x, y) = (X, Y, Z)$ (which is to be interpreted as defined in the preceding section), as well as an alpha value $\alpha_I(x, y)$.

#### 1.1.2.3 Video

Video can be understood as an "image that changes over time", or equivalently, as a function that assigns an image to every point in time. The latter interpretation directly connects video representation to still image representation and can be rephrased into the following definition:

---

[8]The reader should note that any two-dimensional manifold may (locally) be mapped to a plane; therefore this definition naturally extends to more complex two-dimensional shapes.

**Definition 5** *A **video** V is tuple* $(T_V, I_V)$ *where*

- $T_V = [t_{begin}; t_{end}]$ *is a temporal interval*
- $I_V$ *is a function mapping each point* $t \in T_V$ *to an image* $I_V(t)$

## 1.2 Compositing and processing

The previous sections presented mathematical models for audible and visible media. This section will introduce compositing and processing operations that transform media data to achieve a desired effect. The effects can be completely artificial, but are usually modelled after physical effects. The following sections will also introduce formal definitions of these compositing and processing operators.

### 1.2.1 Audio processing

In physical reality, the audio signals emitted by sources need to travel through a medium (or several media) before they can be received by the ear. This may affect the signals in several ways:

1. Delay due to wave propagation latency
2. Attenuation as the wave dissipates into space and loses energy due to friction
3. Frequency-dependent filtering (including delay and attenuation) in a dispersive medium

Furthermore, signals may be reflected by physical objects, and multiple signals may "mix" (we will assume undisturbed superposition[9]). Given a signal function $s_i$ the above effects can mathematically be modelled as:

1  Delay by $\tau$: $s_o(t) = s_i(t - \tau)$

2  Attenuation by $a$: $s_o(t) = a \cdot s_i(t)$

3  Filter with impulse response function[10] $k$: $s_o(t) = (k * s_i)(t)$

4  (Non-linear) transfer by function[11] $f$: $s_o(t) = f(s_i(t))$

---
[9]Air can be modelled as an ideal gas (where the principle of undisturbed superposition of pressure waves holds) as long as the pressure is well below $20Pa$. Since this is roughly the threshold pressure for ear damage, undisturbed superposition can always be assumed in acoustics.

[10]Note that any *linear, time-invariant* effect can be represented as a convolution filter using the effect's impulse response.

[11]We will generally assume the transfer function to be monotonic.

(We could have treated delay and attenuation as special cases of simple impulse response filters). The above set of operations (together with linear combination of signals for superposition) therefore provides a complete compositing algebra for audio signals if the environment (including listener, audio sources) remains static.

If the environment is dynamic (e.g. positions of listener, audio sources or other physical objects are allowed to move), then the requisite parameters $\tau$, $a$, $k$ and $f$ become dependent on time $t$. Assuming that the environmental changes are "slow" (i.e. movement of objects slow compared to audio propagation latency), they can usefully be approximated through linear interpolation between two "key frames" at points in time $t_1$ and $t_2$, i.e:

- Delay: $s_o(t) = s_i \left( t - \left( \frac{t-t_1}{t_2-t_1} \tau_1 + \frac{t_2-t}{t_2-t_1} \tau_2 \right) \right) = s_i \left( \left( \frac{\tau_1 - \tau_2}{t_2 - t_1} + 1 \right) \cdot t + \frac{t_2 \tau_2 - t_1 \tau_1}{t_2 - t_1} \right)$

- Attenuation: $s_o(t) = \left( \frac{t-t_1}{t_2-t_1} a_1 + \frac{t_2-t}{t_2-t_1} a_2 \right) \cdot s_i(t) = \frac{t-t_1}{t_2-t_1} a_1 \cdot s_i(t) + \frac{t_2-t}{t_2-t_1} a_2 \cdot s_i(t)$

- Filter: $s_o(t) = \left( \left( \frac{t-t_1}{t_2-t_1} k_1 + \frac{t_2-t}{t_2-t_1} k_2 \right) * s_i \right)(t) = \frac{t-t_1}{t_2-t_1} (k_1 * s_i)(t) + \frac{t_2-t}{t_2-t_1} (k_2 * s_i)(t)$

- Transfer: $s_o(t) = \left( \left( \frac{t-t_1}{t_2-t_1} k_1 + \frac{t_2-t}{t_2-t_1} k_2 \right) * s_i \right)(t) = \frac{t-t_1}{t_2-t_1} f_1(s_i(t)) + \frac{t_2-t}{t_2-t_1} f_2(s_i(t))$

Since attenuation and filtering are linear operations, they can be exchanged with interpolation, and interpolation requires multiplication with a function linear in $t$. Thus the above operations can be expressed as suitable combinations of:

1' Temporal speedup by $\delta$ and delay by $\tau$: $s_o(t) = s_i(\delta \cdot t - \tau)$

2' Multiplication with envelope function $e : \mathbb{R} \mapsto \mathbb{R}$: $s_o(t) = e(t) \cdot s_i(t)$

3' Filter with impulse response function $k$: $s_o(t) = (k * s_i)(t)$

4' (Non-linear) transfer by function $f$: $s_o(t) = f(s_i(t))$

This set of operations (together with linear combination of signals for superposition) provides a compositing and processing algebra for audio signals that can represent a static environment and approximate a time-dependent environment to arbitrary precision through linear interpolation.

### 1.2.2 Still image processing

The term *image processing* usually has a very broad meaning, encompassing processes that take images (and other data controlling the processes) as input to produce other images or extract features from the original images[12]. For the purposes of multimedia the focus of interest is in processes that generate other images, and the term *image processing* will only be used in this narrow sense:

---
[12][22] pp. 2–3

## 1.2. COMPOSITING AND PROCESSING

**Definition 6** *An **image processing operator** is a function $p$ taking zero or more images $I_1, I_2, \ldots, I_n$ as well as arbitrary other data $D$ as input and produces an image $I = p(I_1, I_2, \ldots, I_n, D)$.*

Image processing operators used in multimedia applications can generally be decomposed into combinations of operators of the following classes (see figure 1.5):

**Point operators** transform each point of an input image individually (and uniformly) to produce a corresponding point of the output image. Typical examples include color transformations such as brightness, contrast or saturation adjustments. Each point operator is uniquely characterized by a single function $f$, that takes a color triplet and an alpha value as parameter, and maps to another color triplet and alpha value.

**Convolution operators** compute each point of the output image by applying a convolution to the input image using a given kernel function. Typical examples include blurring or edge detection filters. Each convolution operator is uniquely characterized by the two-dimensional function used as convolution kernel.

**Geometric operators** translate each point of an input image to a different spatial position but leave its color and alpha value unchanged. Typical examples include scaling and rotation of images, or more generic distortion transformations. Each geometric operator can be characterized through a single function $f : \mathbb{R}^2 \mapsto \mathbb{R}^2$ that describes how the point $f(x,y)$ of the input image maps to the points $(x,y)$ of the output image[13]. In practical applications, geometric operators are usually at least piecewise-smooth and can be approximated sufficiently well using piecewise-linear geometric operators.

**Compositing operators** compute each point of the output image by "compositing" the corresponding points of two input images. Each compositing operator is characterized by a transfer function $f$ that maps color and alpha values of two input images, $(c_A, \alpha_A)$ and $(c_B, \alpha_B)$ to a color and alpha value $(c_O, \alpha_O) = f(c_A, \alpha_A, c_B, \alpha_B)$ of the output image. In other words: each point of the output image uniformly depends on the two points of the input images at the same coordinate location. Examples include the well-known Porter-Duff compositing operators [50], e.g.:

- $A$ **OVER** $B$: represents $A$ layered on top of $B$ and is defined by:
$$c_C = c_A + c_B(1 - \alpha_A)$$
$$\alpha_C = \alpha_A + \alpha_B(1 - \alpha_A)$$
(see also figure 1.6).

- $A$ **IN** $B$: represents $A$ "punched out" by $B$ (i.e. only those parts of $A$ where $B$ is non-transparent) and is defined by:
$$c_C = c_A \alpha_B$$
$$\alpha_C = \alpha_A \alpha_B$$

---
[13]In other words, the function $f$ determines the *preimage* of every point.

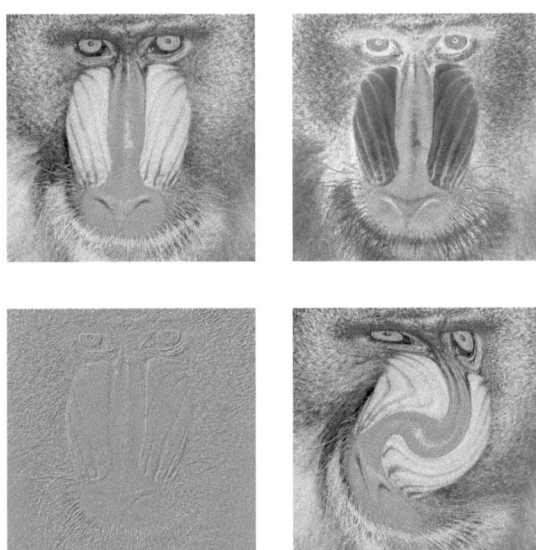

*Top left: Original image. Top right: Point operator (color inversion). Bottom left: Convolution operator (edge detection filter). Bottom right: Geometric operator (spiral distortion).*

Figure 1.5: Examples for unary image processing operators.

- $A$ ATOP $B$: represents $A$ composited over $B$ where $B$ is non-transparent and is defined by:
$$c_C = c_A \alpha_B + c_B(1 - \alpha_A)$$
$$\alpha_C = \alpha_A \alpha_B + \alpha_B(1 - \alpha_A) = \alpha_B$$

While the four classes of operators listed above are certainly not "universal" (in that they can construct arbitrary image processing operators), but they still represent a vast superset of what practical graphics systems offer.

## 1.3 Representation in digital systems

Audible and visual media were introduced in sections 1.1.1 and 1.1.2 respectively as functions mapping from continuous time and/or spatial domains into continuous intensity domains, e.g. a video is conceptually a function of the domain $\mathbb{R}^3 \mapsto [0;1]^4$. The set of all possible media is obviously uncountable, digital systems can however only represent and process a finite subset of this domain. This section will discuss media representations that cover or approximate an "interesting" subset of this domain, and which are therefore used in digital systems.

## 1.3. REPRESENTATION IN DIGITAL SYSTEMS 21

*Top left: first image. Top right: second image. Bottom left: alpha channel of second image (black: fully transparent, white: fully opaque). Bottom right: Composition of first and second image (OVER operator)*

Figure 1.6: Example of compositing operators.

### 1.3.1 Techniques

#### 1.3.1.1 Quantizing

*Quantizing* is the process of expressing some infinitely variable quantity by discrete or stepped values[14]. The range of representable values is partitioned into a finite set of (disjoint) intervals, and a representative is chosen for each interval (that representative must be a member of the interval). Each interval is encoded, usually using a unique integral number. Instead of the original value of the quantity, this encoding is used for storage and/or processing instead. The inverse process, dequantizing, replaces the coded integers by the representatives of the corresponding intervals. The difference between the original quantity and the representative is called *quantizing error*. Note that dequantizing and subsequent quantizing is idempotent by this definition.

Intervals do not have to be evenly sized, and the representatives do not necessarily have to be placed at the center of the interval they represent (cf. figure 1.7). A quantizing is said to be *linear* iff all intervals are of the same size and for each interval the representative value depends linearly on the integral number

---
[14][69] pp. 56ff

# 22  CHAPTER 1. MULTIMEDIA REPRESENTATION AND PROCESSING

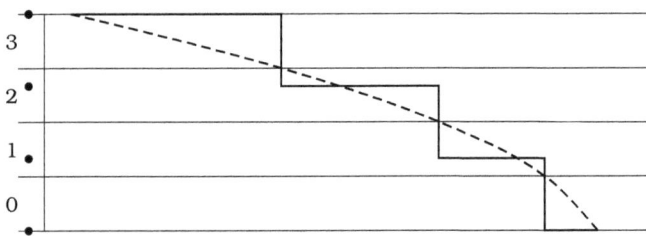

The value range of a continuous quantity is partitioned into four disjoint intervals (horizontal lines) with a representative value (dots on the left) and code (number on the left) chosen for each interval. A continuous signal (dashed line) that represents the quantity changing over time is quantized by determining and encoding the interval the value is in. Dequantizing is performed by replacing the coded intervals with the chosen representative (solid line).

Figure 1.7: Quantizing a continuous quantity

used for encoding. Linear quantizing has the important property that arithmetic operations on the values of the quantity can be approximated by corresponding arithmetic operations on the encoding values used to represent the intervals. Therefore, linear quantizing is preferentially used where the data needs to be processed further. (Note that the quantizing scheme illustrated in figure 1.7 can be considered linear).

Non-linear quantizing allows to adapt the quantizing step sizes to the problem domain, choosing smaller intervals where more precision is required. This is especially common where the *relative* quantizing error needs to be bounded: Linearly quantizing the interval $[0;1]$ with $N$ quantizing steps results in an average absolute quantizing error of (at best) $1/2N$ which means that the relative quantizing error exceeds one for values below $1/2N$. Choosing intervals $[0; \alpha^{1-N})$, $[\alpha^{1-N}; \alpha^{2-N})$, ..., $[\alpha^{-1}, 1]$ with $1 < \alpha < 2$, then relative quantizing error remains bounded by $1/(1-\alpha)$ for all values larger than $\alpha^{1-N}$.

Note that non-linear quantizing can always be interpreted as a non-linear monotonic transformation followed by linear quantizing.

### 1.3.1.2  Sampling and interpolation

*Sampling* is the process of taking values of a function (defined over the continuum) at a countable (usually finite) number of points (cf. figure 1.8). Usually, the sampling points are chosen equidistant to cover an interval of interest, however this is not of necessity. The inverse process, constructing a function that connects a given set of samples, is called *interpolation*. Different interpolation strategies may be used depending on the application domain.

## 1.3. REPRESENTATION IN DIGITAL SYSTEMS

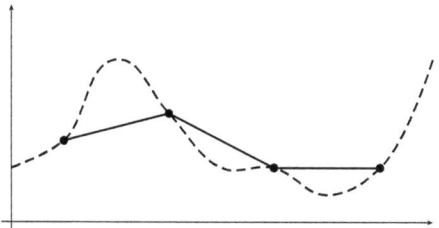

*A function (dashed line) defined over a continuum is sampled at discrete points (solid circles). The finite set of samples can then be interpolated to produce an approximation of the original function (solid line)*

Figure 1.8: Sampling a continuous signal

Let $S$ be a specific chosen sampling strategy, i.e. it maps the function $f$ to the set of sample point/value pairs $S(f)$[15]. Let $I$ be an interpolation strategy, i.e. it maps a set of samples $s$ to an interpolating function $I(s)$. We demand that the interpolating function $I(s)$ "passes through" the samples, as a consequence we have $S(I(s)) = s$.

Given a function $f$, the combination of sampling and interpolation constructs an approximation $I(S(f))$ to the function $f$, but generally $I(S(f)) \neq f$ (i.e. the original function $f$ cannot be reconstructed after sampling). Furthermore, the function $I(S(f))$ will generally not be the best approximation to $f$[16]. Sampling is therefore usually combined with a pre-sampling filter $F$, where $F$ maps a function $f$ to another function $F(f)$ such that $I(S(F(f)))$ is the best approximation to $f$. The filter $F$ is generally referred to as *anti-aliasing filter*. Examples for domain-specific interpolation and anti-aliasing filters will be given in sections 1.3.2.1 and 1.3.4.1.

Combining *sampling* and *quantizing* (cf. section 1.3.1.1) allows an approximation to a continuous function to be represented using a finite number of bits.

### 1.3.2 Audio representation

Audio signals were introduced as functions of the domain $\mathbb{R} \mapsto \mathbb{R}$. Digital systems can only represent the subset of computable functions. The most obvious representational choice is a mathematical term. This representation can be used to describe simple synthesis, e.g. using frequency modulation:

---

[15]Essentially, $S(f)$ is the restriction of $f$ to the set of sample points.
[16]as measured by a problem-specific metric

$$a_c(t)\sin\left(2\pi f_c t + a_m(t)\sin(2\pi f_m t)\right))$$

with piece-wise linear envelope functions $a_c$ and $a_m$. This approach can be used to model the behavior of analog synthesizers in digital systems and is therefore sometimes applied in digital music applications. Though unwieldy, more complex audio synthesis can formally be expressed in a similar fashion: e.g. waveguide synthesis [57] can be regarded as a combination of PCM audio (see section 1.3.2.1 below), elementary functions representable as terms, and the audio compositing operators discussed in section 1.2.1.

PCM audio (discussed in the following sections) can be considered an important special case, as it can formally be interpreted as a continuous signal interpolating a set of given sample points, and the interpolating function is computable.

#### 1.3.2.1 PCM audio

We assume a representation of an audio signal that is sampled at temporally equidistant points:

**Definition 7** *A **time-discrete audio signal** $A'$ is a triple $(T_{A'}, s_{A'}, n_{A'})$ where*

- *$n_{A'}$ is the number of sample points*

- *$T_{A'} = \{t_{begin}, t_{begin} + \frac{1}{n_{A'}}\Delta t, t_{begin} + \frac{2}{n_{A'}}\Delta t, \ldots, t_{begin} + \frac{n-1}{n_{A'}}\Delta t\}$ (with $\Delta t = t_{end} - t_{begin}$) is a set of equidistant points in time within the interval $[t_{begin}; t_{end})$*

- *$s_{A'} : t_{A'} \to \mathbb{R}$ is a function mapping each point in time $t$ to the pressure in a medium $s_{A'}(t)$ at that point in time*

(For convenience, the sample at temporal location $t_{begin} + \frac{k}{n_{A'}}\Delta t$ will often be identified by $k$ alone instead of the corresponding point in time).

Audio signals that are both time-discretized and quantized will be referred to as *PCM audio signals*. Depending on the application's intent linear or non-linear quantizing may be preferable (cf. section 1.3.1.1): Linear quantizing is preferentially used for audio processing as arithmetic operations on the sample values can be approximated by arithmetic operations on the ordinal number chosen to represent the quantizing intervals. Assuming $N$ quantizing steps, the average absolute quantizing error is $2/N$ – this means that for $N = 255$ the quantizing error exceeds the signal amplitude at $-42dB$ (or for $N = 65535$ at $-90dB$). Thus for linear quantizing the number $N$ of quantizing steps imposes an upper bound on the representable dynamic range. Note that the quantizing error is uniform across the whole value range, thus superposition of high-amplitude signals does not adversely affect low-amplitude signals.

## 1.3. REPRESENTATION IN DIGITAL SYSTEMS

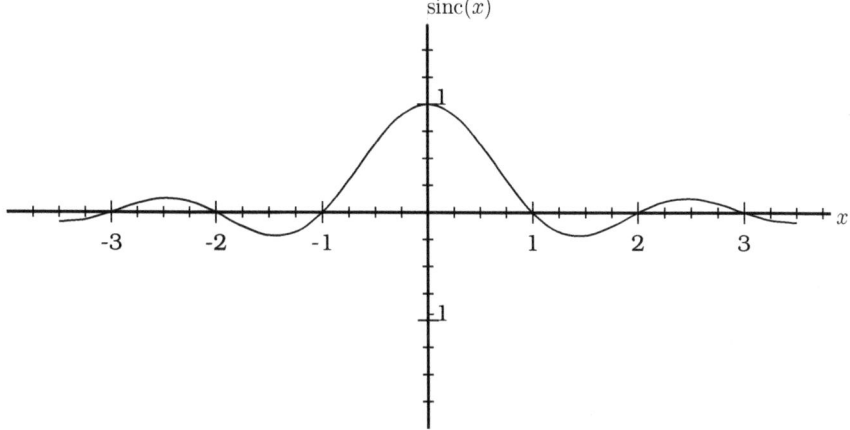

Figure 1.9: Plot of the base function $x \mapsto \mathrm{sinc}(x)$ used for band-limited interpolation of audio signals

Non-linear quantizing schemes are used to encode a larger dynamic range into the same number of bits per sample than would be possible with linear quantizing. A popular example includes $\mu$-law encoding, which can be understood as linear quantizing after transforming the input signal using

$$F(x) = \mathrm{sgn}(x)\frac{\ln(1+\mu|x|)}{\ln(1+\mu)}$$

with $\mu = 255$ typically. Assuming 255 quantizing steps, the quantizing error exceeds the signal amplitude at $-75dB$. Note however that this increase in dynamic range comes at the cost of a higher absolute quantizing error, thus superposition with high-amplitude signals leads to heavy distortion or complete masking of low-amplitude signals.

Band-limited signals of band-width $w$ can be reconstructed perfectly if sampled at a rate of greater than $2w$ (see next section). Since human hearing is limited to the range of about $20Hz$ to $20kHz$ perfect acoustic fidelity can therefore be achieved with 40000 samples per second. In practice, sample rates lower than $40kHz$ are in use to reduce the number of samples (and thus bits) required to represent the signal (thus they are more suitable for *storage* or *transmission*), while sampling frequencies higher than $40kHz$ simplify construction of digital filters (thus they are more suitable for *processing*, see next section).

### 1.3.2.2 PCM audio interpolation

Interpolation of time-discrete audio signals $A_d$ (and PCM audio signals as a further specialization) to continuous signals $A_c$ will be performed using the Whittaker-Shannon interpolation formula:

$$s_{A_c}(t) = \sum_{t' \in T_{A_d}} s_{A_d}(t') \operatorname{sinc}\left(\frac{t-t'}{\Delta t}\right)$$

The interpolation can be visualized as a superposition of scaled versions of the sinc basis function (cf. figure 1.9) placed at node points. Since sinc vanishes to zero at other node points, the requirements of section 1.3.1.2 are met. Furthermore, sinc is ideally band-limited, therefore the resulting interpolant is ideally band-limited as well – the interpolating function is the *unique* band-limited continuation of the given sample data.

Given an arbitrary function $f$, the best interpolation of $f$ (in $L^2$) through a band-limited function is given by $f * \operatorname{sinc}$. Therefore, the filter function to be applied before sampling an audio signal takes a particularly simple form: it is the *ideal low-pass filter* (characterized by the impulse response sinc itself).

The theoretically correct procedure for constructing a time-discrete representation of an audio signal is therefore to apply an ideal low-pass filter before sampling the function. This can however usually not be performed in practice, as the ideal low-pass filter a) does not have finite impulse response (and would for computations in the time domain therefore require an infinite number of operations) and b) is non-causal (it affects information infinitely backwards in time).

Practical low-pass filter representations can therefore only approximate the desired ideal filter. The most widely used method to construct such approximating filters is to "window" the ideal low-pass filter response function sinc such that it smoothly vanishes to zero at a finite distance from the origin. This filter design approach is used in a variety of applications, e.g. for band-limited resampling [58], other examples include the band-pass filters used in MP3 coding (see section 1.4.1.3).

The ideal low-pass filter completely eliminates all signals in the "stop-band", leaving all other signals (the "pass-band") unchanged. A practical filter only attenuates "stop-band" signals below a designed threshold, signals in the "pass-band" are "mostly" unchanged, and exhibits intermediate behavior in a "transition band". The width of the transition band directly affects the computational complexity of numerical filter implementations, with larger available bands leading to computationally simpler filters. For example, assuming that a band-width of $40kHz$ is required to losslessly represent acoustic signals, a sample rate of $44.1kHz$ allows a transition band of only $4.1kHz$ width, compared to $8kHz$ for $48kHz$ sample rate. Thus higher sampling frequencies allow computationally less expensive filter designs and are therefore prefered for audio processing.

## 1.3.3 Color representation

While CIE XYZ allows an objective representation of color, other *color models* are often more convenient:

**Definition 8** *A* **color model** *$C$ is a tuple $(colors_C, map_C)$ where*

- *$colors_C$ is a closed subset of $\mathbb{R}^3$*
- *$map_C$ is a continuous injective mapping $colors_C \to [0, \infty)^3$*

$colors_C$ *is the range of colors representable within color model $C$ and $map_C$ maps every triplet of values in $colors_C$ to a* **color** *(as per definition 3).*

Note that while sometimes used interchangeably with *color model* both in literature and by practitioners, the term *color space* in the strict sense means only the set of colors representable within a color model: Since each color model implicitly defines a color space this slight inaccuracy may be excusable. Typically, color models are normalized such that $colors_C = [0; 1]^3$, however the definition is intentionally generic (e.g. $YP_rP_b$ uses the color range $[0; 1] \times [-0.5; +0.5] \times [-0.5; +0.5]$). Note that for storage in digital systems the color triplet components are typically quantized.

**Definition 9** *The color model $C$ is called* **(affine-)linear** *iff $map_C$ is (affine-)linear.*

Trivially CIE XYZ itself can be regarded as a linear color model. A special class of non-linear color models is of particular interest:

**Definition 10** *The color model $C$ is called* **gamma-corrected** *iff there exist*

- *continuous monotonic scalar functions $\gamma_a$, $\gamma_b$, $\gamma_c$*
- *an affine-linear color model $C'$ (with suitable color range $colors_{C'}$) such that $map_C(a, b, c) = map_{C'}(\gamma_a(a), \gamma_b(b), \gamma_c(c))$*

The term *gamma* refers to the fact that $\gamma$ is often given through a power law of the form $x \mapsto x^\gamma$. Gamma-corrected color models are used extensively in practice for the following reasons:

- Devices for physical reproduction or acquisition of colors (e.g. CRTs, CCDs) exhibit a non-linear relationship between light intensity and signal.
- The photoreceptor cells become less sensitive to small intensity differences at higher intensities, thus non-linear quantizing is often employed[17].

---

[17] This can be understood as a combination of non-linear transformation and linear quantizing.

# CHAPTER 1. MULTIMEDIA REPRESENTATION AND PROCESSING

The purposes of alternate color models in practical use in the field of multimedia can roughly be classified into

- **physical**: Each component of the triplet corresponds to one physically emittable color (e.g. "red", "green" and "blue").
- **decorrelation**: Each component corresponds to a separate psycho-visual "feature" (e.g. "brightness" and color tone).
- **compositing**: The topology of the color model is suitable for logical operations on colors to be expressed as simple mathematical operations on the vector space (i.e. blending of colors as interpolation along a straight line).

Transformations between two color models can in principle be defined in a straight-forward fashion (as transformations into CIE XYZ and back). Note that the definition of color model above may limit the value range of the three coefficients so that in many cases no perfect one-to-one mapping between different color models is possible, as some colors from the source color model are unrepresentable in the target color model. In practice transformations will therefore either

- map "unrepresentable" colors to the "closest" matching color, preserving *absolute* colors where an exact match can be found
- map all colors from the source color model to a "close" matching color in the target color model, preserving *differences* between colors, but subtly distorting colors overall

Note that the second strategy can only be usefully applied if the full source color range is known before the transformation must be applied; it may therefore be suitable for still images, but is generally unsuitable for motion pictures as the color range of "future" images is most certainly unknown before transforming the first image. While applications may provide their own strategy on a case-by-case basis [61] we will therefore assume the first strategy as a "safe default".

The following sections will give examples for different classes of color models that are referenced later. Note that for storage and processing in digital systems, the continuous-valued triplet components will generally be quantized.

## 1.3.3.1 Physical color models

Physical reproduction of color is usually achieved using a "red", "green" and a "blue" primary color. This naturally leads to the definition of a (family of) "RGB" color models. One example of particular interest is CIE RGB, defined through

$$\begin{pmatrix} X \\ Y \\ Z \end{pmatrix} = map_{CIE\_RGB} \begin{pmatrix} R \\ G \\ B \end{pmatrix} = \frac{1}{0.17697} \begin{pmatrix} 0.49 & 0.31 & 0.20 \\ 0.17697 & 0.81240 & 0.01063 \\ 0.00 & 0.01 & 0.99 \end{pmatrix} \begin{pmatrix} R \\ G \\ B \end{pmatrix}$$

## 1.3. REPRESENTATION IN DIGITAL SYSTEMS

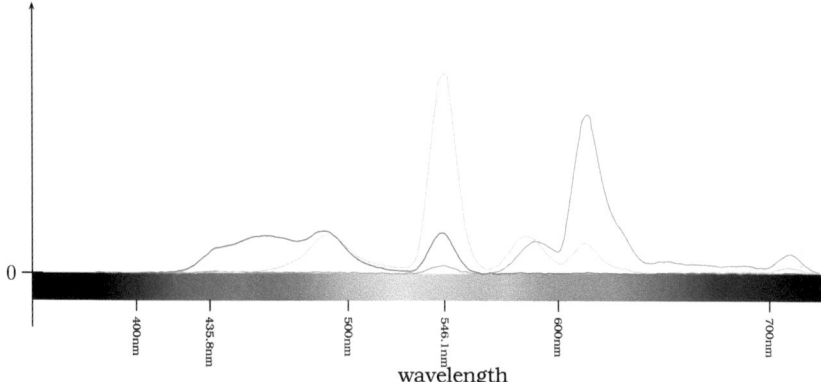

*The curves were obtained from an Asus model LC5800 laptop by optical spectrometry: The light emitted by red, green and blue phosphors was spectrally split up using a grid mono-chromator, and light intensities at individual wavelengths were determined using a silicon diode as photoreceptor.*

Figure 1.10: Emission spectra of a typical liquid crystal display

In this color model, the components of the $(R, G, B)$ triplet correspond to the intensity of monochromatic light at wavelengths $700nm$ ($R$), $546.1nm$ ($G$) and $435.8nm$ ($B$) (see also figure 1.3). CIE RGB is notable as its three primary colors were the ones used in experiments that led to the definition of both CIE XYZ and CIE RGB.

Device-dependent RGB color models can be defined for capture and display devices. Typical CRT displays have emission spectra comparable to those shown in figure 1.10 (the reader should take note that the data shown in the diagram is in fact sufficient to derive the mathematical relationship to CIE XYZ) and provides an example of a *gamma-corrected* RGB color model (as the signal response of CRTs in non-linear).

Several device-independent color spaces were created that mimic the properties of existing device-dependent color spaces. One such example is *sRGB*: linear sRGB is defined through:

$$\begin{pmatrix} X \\ Y \\ Z \end{pmatrix} = map_{linear\_sRGB} \begin{pmatrix} r \\ g \\ b \end{pmatrix} = \frac{1}{0.17697} \begin{pmatrix} 0.4124 & 0.3576 & 0.1805 \\ 0.2126 & 0.7152 & 0.0722 \\ 0.0193 & 0.1192 & 0.9505 \end{pmatrix} \begin{pmatrix} r \\ g \\ b \end{pmatrix}$$

while *gamma-corrected sRGB* (or simply sRGB) is related to linear sRGB through the transfer function $\gamma$ defined by

$$\gamma(c) = \begin{Bmatrix} 12.92c, & c \leq 0.0031308 \\ (1 + 0.055)c^{1/2.4} - 0.055, & c > 0.0031308 \end{Bmatrix}$$

Other examples of RGB color models include e.g. PAL RGB which is specified as using the primary colors

| primary | x | y | Y |
|---|---|---|---|
| red | 0.6400 | 0.3300 | 0.222021 |
| green | 0.2900 | 0.6000 | 0.706645 |
| blue | 0.1500 | 0.0600 | 0.071334 |

and the gamma transfer function

$$\gamma_{ITU\_Rec.709}(c) = \begin{cases} \frac{c}{0.018}\left((1.099*0.018^{0.45})-0.099\right) & \text{for } c \leq 0.018 \\ \left((1.099*c^{0.45})-0.099\right) & \text{for } c > 0.018 \end{cases}$$

The color models of input and output devices are in practice usually specified through an ICC color profile which in turn references CIE XYZ as described above or CIE L*a*b as described below.

### 1.3.3.2 Decorrelation color models

For digital video color models that decorrelate perceived "brightness" and "chromaticity" of a color are desirable; the motivation is twofold: First, the human eye is more sensitive to spatial differences in brightness than chromaticity; second, physical objects often have uniform chromaticity while brightness varies due to illumination effects. The most common (family of) color models used is $Y'P_bP_r$[18] which is defined relative to (gamma-corrected) PAL RGB through

$$\begin{pmatrix} Y' \\ P_b \\ P_r \end{pmatrix} = \begin{pmatrix} 0.299 & 0.587 & 0.144 \\ -0.168736 & -0.331264 & 0.5 \\ 0.5 & -0.418688 & -0.081312 \end{pmatrix} \begin{pmatrix} R' \\ G' \\ B' \end{pmatrix}$$

Note that $Y'P_bP_r$ is defined for $[0;1] \times [-0.5;+0.5] \times [-0.5;+0.5]$. Another closely related color model is $Y'C_bC_r$ defined through

$$\begin{pmatrix} Y' \\ C_b \\ C_r \end{pmatrix} = 255 \begin{pmatrix} Y' \\ P_b \\ P_r \end{pmatrix} + \begin{pmatrix} 0 \\ 128 \\ 128 \end{pmatrix}$$

$Y'C_bC_r$ was defined for digital PAL television but it is used in many other digital video applications as well; it remaps the value range of $Y'P_bP_r$ into $[0;255]^3$, due to its use in television the true value range of $Y'C_bC_r$ is however limited to $[16;235] \times [16;240] \times [16;240]$ (codes outside this range are used for control purposes)[19].

---

[18]Note that the "luma" value $Y'$ is often incorrectly mixed up with the "luminance" $Y$ from CIE XYZ or CIE xyY; while both serve a similar purpose they are not to be confused as $Y$ measures linear brightness whereas $Y'$ measures a pseudo-brightness derived from gamma-corrected RGB: See [39] for more details.

[19]However, very few implementations outside the field of television actually care about these ranges.

## 1.3. REPRESENTATION IN DIGITAL SYSTEMS

#### 1.3.3.3 Compositing color models

For image compositing, logical operations on colors have to be expressed in terms of algebraic operations on the underlying color model; the result of these algebraic operations should closely match experience of the physical world (e.g. blending between a "red" and "yellow" color should result in some shade of "orange", but no blue).

All linear color models share the property that blending between colors can usefully be expressed as interpolation along a straight line between the vectors representing the two colors. However, linear color spaces are not perceptually uniform – distance between two vectors (in the usual $\mathbb{R}^3$ metric) does not correspond to perceived difference of colors (this is most visible when constructing color gradients).

One such perceptually uniform color space is CIE L*a*b; it is related to CIE XYZ through

$$L = 116\sqrt[3]{Y} - 16$$
$$a = 500 \left( \sqrt[3]{\frac{X}{0.95}} - \sqrt[3]{Y} \right)$$
$$b = 200 \left( \sqrt[3]{Y} - \sqrt[3]{\frac{Z}{1.09}} \right)$$

CIE L*a*b is also often used as an intermediate color model for conversion. Its perceptual uniformity is useful in finding a "closest matching color" when no exact color match can be found in the target color space. Generally perceptual uniformity is more important for artistic than technical purposes as most physical effects affecting color are better expressed in linear color spaces.

### 1.3.4 Image representation

Images were introduced as functions mapping a rectangular area of $\mathbb{R}^2$ to the set of color (and alpha) values. Digital systems can only represent the subset of computable functions out of this domain. The most obvious representational choice, a mathematical term used to describe the dependence of the color of a point from its position, is frequently used in computer graphics to describe source images (such as color gradients) for compositing operations (cf. sections 1.2.2). Obviously, applying computable image processing operators to computable images provides another way to represent an image.

Note that representation using one of the color models discussed in section 1.3.3 may be more convenient for this purpose than the XYZ color model defined in section 1.1.2.1.

*Rastered images* as discussed in the following sections are an important special case of computable image representations.

## 1.3.4.1 Rastered images

We assume a representation of an image that is sampled at node points of a rectangular, regular grid:

**Definition 11** *A* **space-discretized image** *$I$ is a quintuple $(C_I, w_I, h_I, c_I, \alpha_I)$ where*

- *$C_I$ is a color model*
- *$w_I$ and $h_I$ are positive integers*
- *$c_I$ is a mapping $\{0, 1, 2, \ldots, w_I - 1\} \times \{0, 1, 2, \ldots, h_I - 1\} \to \mathbb{R}^3$*
- *$\alpha_I$ is a mapping $\{0, 1, 2, \ldots, w_I - 1\} \times \{0, 1, 2, \ldots, h_I - 1\} \to [0; 1]$*

The definition is to be interpreted in a similar way as definition 4: The function $c_I$ maps every grid point to a triplet of values, which must be interpreted using $C_I$ as color model. Note that this definition uses the same grid for all components of the color triplet and the alpha channel; in practice image formats with different grids for each component are in use, we will however refrain from presenting the (obvious) formalization of this concept.

Images that are both *space-discretized* and *quantized* will be referred to as *rastered images*. Linear quantization in combination with a linear color model is preferentially used for image processing, while storage and transmission of image data almost always uses non-linear quantization[20]. Most notably – at the time of this writing – display systems exhibit non-linear response with gamma values typically in the range of 1.6 to 2.4.

## 1.3.4.2 Rastered image interpolation

Continuation of a space-discretized image $I$ to a space-continuous image $I'$ is realized through:

$$c_{I'}(x, y) = map_{C_I}(c_I(\lfloor x \rfloor, \lfloor y \rfloor))$$

i.e. each color sample is continuated to a $1 \times 1$ square area. This interpolation strategy models the behavior of physical display devices that use uniformly colored cells as elementary "tiles" to compose images. The ideal pre-sampling filter is given by convolution with the kernel function

$$\chi_{[0;1) \times [0;1)}(x, y) = \begin{cases} 1 & \text{iff } 0 \leq x < 1 \text{ and } 0 \leq y < 1 \\ 0 & \text{otherwise} \end{cases}$$

This filter "averages" the color intensities of all points in a unit square. The reader may wish to convince himself that the above interpolation and filtering strategy leads to *bilinear interpolation* if resampling a rastered image using a

---

[20] or, equivalently, linear quantization with a non-linear color model

## 1.3. REPRESENTATION IN DIGITAL SYSTEMS

displaced or rotated grid. More generally, it leads to desirable results for any resampling operation that uses a grid with equal or more coarse granularity than the rastered source image. For magnification (i.e. using a more fine-grained sampling grid) other filters may be desired for aesthetic reasons, e.g. to slightly "blur" the image rather than showing large tiles.

### 1.3.4.3 Rastered image processing

Of the image processing operators discussed in section 1.2.2, point and compositing operators can be rephrased in terms of rastered images in a straightforward fashion: They affect all points of the image individually and uniformly, so they can be applied to the interpolated image or to the sample values with subsequent interpolation, with the same result. Formulation of convolution operators must generally take into account the anti-aliasing filter required before sampling – this can be accomplished by factoring anti-aliasing directly into the convolution kernel.

Geometric processing operators on the other hand are more complex – essentially geometric distortion can be interpreted as resampling the image using a geometrically distorted grid. Proper filtering needs to be applied before sampling the image, however construction of a computable filter becomes non-trivial in the general case. In practical applications it is therefore customary to split geometric processing operators into piecewise affine-linear transformations (for affine-linear grids the filter kernel just becomes an affine-linear image of the "unity square" ideal filter from the previous section).

The sampling filter discussed in the previous section also leads to a precise definition for the rasterization of geometric shapes (where a point covered by the shape is considered to have full opacity $\alpha = 1$ and a point not covered is considered to be fully transparent $\alpha = 0$): The $\alpha$ value of a sample point corresponds to the coverage ratio of the corresponding rectangular area[21]. The image compositing operators described in section 1.2.2 are defined to use the alpha channel such that compositing of pre-rastered geometric figures with other images approximates the compositing of exact geometric figures.

Alpha values to represent pixel coverage for geometric shapes are rarely determined using the "analytical" definition given above as the computation turns out to be prohibitively expensive for complex shapes. Typical applications use "super-sampling", i.e. the shape is sampled at a higher resolution with a simple inside/outside test for each sample point; sample points comprising a pixel are then averaged to form an alpha value. This approach inherently leads to quantized alpha values, with the number of super-samples limiting the number of possible quantization steps (non-uniform weighting of super-samples can however considerably reduce the number of super-samples required, cf. [35]).

---

[21] Note that Porter, Duff [50] also use sub-pixel geometry in deriving the compositing operators.

## 1.3.5 Video representation

Video was introduced as time- and space-continuous. Out of the uncountable set of possible video sequences only a computable subset can be represented in digital systems. "True" time-continuous video can be represented e.g. as keyframe animation (cf. [59]): a continuously variable quantity that determines appearance of objects is interpolated between keyframes. Most commonly however *time-discretized* video representations are used.

### 1.3.5.1 Time-discretized video

We will define time-discretized video as a sequence of still images that are shown for a fixed duration of time each:

**Definition 12** *A* **time-discretized video** $V'$ *is tuple* $(T_{V'}, I_{V'})$ *where*

- $T_{V'} = \{t_0, t_1, t_2, \ldots, t_n\}$ *is a set of discrete points in time (we will assume $t_{k-1} < t_k$)*

- $I_{V'}$ *is a function mapping each point* $t \in T_{V'}$ *to an image* $I_{V'}(t)$

Interpolation of a time-discretized video $V'$ to a time-continuous video $V$ is achieved by:

$$I_V(t) = I_{V'}(t_n) \text{ for } t \in [t_n; t_{n+1})$$

The corresponding temporal anti-aliasing filter averages images over the temporal interval where the corresponding still image is to be shown. Note that the definition does not demand that the points $t_k$ are equidistant in time, i.e. the definition allows for variable frame rates.

# 1.4 Compressed representations

This section describes what is commonly referred to as media data "compression", with the common meaning of space savings achieved relative to time- and space-discretized, quantized representations using a fixed number of bits per sample. In the sense of the terminology introduced in section 1.3, it is of course just another kind of computable media representation. Note that in this broad sense, almost *any* computable representation could be considered "compressed", however we want the term to be understood in a more narrow sense, namely representations that

- interpret quantized samples of time- and space-discretized media as "symbols", possibly after performing decorrelation transformations

## 1.4. COMPRESSED REPRESENTATIONS

- encode these symbols into a bitstream, reducing redundancy between symbols (e.g. entropy, dictionary or run-length encodings) to find a more space-efficient encoding

We will frequently distinguish "lossless" and "lossy" representations: "Lossless" means that a given set of sample values can be reconstructed perfectly from an encoding, while "lossy" means that the content may be "slightly" distorted with respect to the original after reconstruction. Strictly speaking, lossless or lossy encoding is a property of the *algorithms* used to transform media data into and out of a compressed representation – but *representations* may not always be suitable for both faithful and lossy encoding as required transformations may not be amenable to precise fixed-point realizations.

Compression techniques make use of the following types of redundancies found in (sampled) media data:

- *Spatial*: Samples corresponding to different but close spatial locations are correlated. For example, a point audio source sampled at two different points in space will produce similar signals (differing only in filtering and temporal shift). Similarly, close points of the same image are correlated as they are likely to correspond to different spots of the same physical object.

- *Temporal*: Samples corresponding to different but close temporal locations are correlated. For example, audio signals tend to be "smooth"[22], thus temporally close samples will likely have similar values. Similarly, temporally close images of a video likely share elements (e.g. unchanging background, objects moving in the scene over time).

- *Spectral*: Audio signals tend to be "repetitive" as they are physically the result of oscillations. As a result, the spectrum tends to be slow-changing and "sparse" (i.e. intensity spikes are concentrated in few narrow frequency ranges).

### 1.4.1 Audio compression

*Spatial redundancy* means that samples corresponding to different channels but "close" points in time are likely similar. This can be exploited by "joint" encoding of multiple channels, i.e. the signals for individual channels are encoded "as a whole" (instead of coding each signal individually). Techniques to achieve this are separate encoding of different audio sources with designated mix-down into individual channels, or decorrelation by encoding e.g. sum and difference of two signals (instead of the original audio signals).

*Temporal redundancy* implies that a sample value can be "estimated" with good accuracy from the surrounding or preceding sample values. This is exploited in "time-domain" representations by predictive coding. *Spectral redun-*

---
[22] This is a direct side-effect of applying a low-pass filter before sampling.

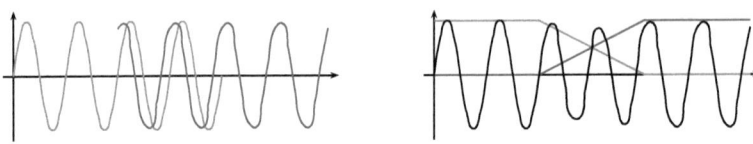

Left: Two signals that cannot be joined smoothly at any single point in the overlap of their definition ranges due to phase differences. Right: The two signals joined smoothly by tapering both functions to zero towards the borders of their definition ranges. This is accomplished through multiplication with window functions that decay to zero towards at the borders.

Figure 1.11: Smoothing transition between signals

*dancy* means that audio signals can usefully be represented as linear combination of periodic functions, with representation of the functions requiring fewer values than the number of samples. (Note that temporal and spectral redundancy are strongly correlated phenomena.)

Reduction of temporal and spectral redundancy is achieved by "joint" encoding of samples corresponding to different points in time. This is usually achieved by forming a "block" of a temporal range of samples, which then becomes the smallest unit of data that can be processed individually. Increasing the block size generally allows more efficient encoding, though at diminishing returns. At the same time, larger blocks result in compression delay (which may be prohibitive in real-time audio processing systems) as all samples comprising the block need to be recorded before transformation can begin.

Since lossless audio compressors faithfully reproduce given sample data, it is implicitly guaranteed that signals are joined smoothly at block boundaries. These transitions pose a problem however for lossy audio compressors, as there is no a priori guarantee that the reconstructed signal remains "smooth" at transitions. This problem is addressed by partial overlapping of adjacent blocks. The samples in the overlapped region are coded in both blocks, and during decompression the functions are "weighted" in the overlapping range (cf. figure 1.11). Any compression schemes where individual samples cannot be reconstructed from a single frame alone will be referred to as *temporal compression*.

### 1.4.1.1 ADPCM

ADPCM (Adaptive Differential Pulse Code Modulation) designates a class of time-domain audio encodings that are based on the idea of a "stateful predictor". It allows each sample to be encoded using a fixed number of bits that is however significantly lower than the number of bits as would be required to represent the full dynamic range of sample values.

## 1.4. COMPRESSED REPRESENTATIONS

In ADPCM the predictor produces an estimate of the value range for the next sample. Based on this prediction, "quantization steps" are chosen accordingly. To be useful, the quantization steps must be small in number (to be representable by few bits), yet must cover the range of values sufficiently well (to avoid heavy signal distortions that occur if the true value is outside of the range chosen by the predictor). The sample value is then quantized and the predictor state is updated to produce a new estimate for the next sample.

Knowledge of the predictor state is essential in interpreting an ADPCM data bitstream. Thus, if random access to the audio data (or resilience agaìnts data loss) is required, the predictor state must periodically be saved or transmitted as well. For the purposes of storage or transmission it is therefore customary to form self-contained frames by first encoding the initial predictor state, followed by a fixed number of coded samples.

Since ADPCM encoding operates on one sample at a time, it adds virtually no latency to audio data processing and is therefore especially popular in application areas with tight real-time bounds. Examples for this class of encodings include ITU G.726 used in many IP-based telephony applications; G.726 specifies multiple alternative encodings using between 2 and 5 bits per sample, IP telephony applications typically use 4 bits per sample.

### 1.4.1.2 FLAC

The FLAC (Free Lossless Audio Codec) format is a "time-domain" audio signal representation with the explicit intent of allowing *lossless* compression of PCM audio [19]. It supports joint encoding of multi-channel audio through decorrelation transformations, after which each of the resulting signals are encoded individually.

Each PCM audio signal is partitioned into blocks (usually in the range of 2048-6144 samples), and each block of samples is encoded individually. The samples in each block are represented as *residual* errors relative to a *model* used as predictor. FLAC supports several different models as predictor, these include

- "zero" predictor (referred to as "verbatim" mode in the FLAC specification): Each sample is effectively predicted as "zero" by this model, the "residual" error to be encoded thus equals the sample value.

- "linear" predictor: Each sample value is predicted as a linear combination of $n$ previous sample and residual values.

Several other models (including a FIR variant of the linear predictor that does not use previous residuals) are supported, and more may be added in future revisions. The residual errors are expected to be concentrated around zero, and FLAC uses a variable length encoding (Rice codes[23]) for residual values, utilizing this statistical property for data reduction.

---
[23] Also known as Golomb-Rice codes, see [21].

38     CHAPTER 1. MULTIMEDIA REPRESENTATION AND PROCESSING

#### 1.4.1.3  MPEG Audio

MPEG audio [28] is a lossy compressed representation for PCM audio data. The basic ideas of the representation will be sketched out here, for a more thorough introduction refer to e.g. [46].

The fundamental idea of representation in MPEG audio is to split the input signal into multiple "subbands", and to treat each subband signal individually afterwards. Consider a signal function $s$ that is band-limited with upper frequency $2n$ (i.e. its Fourier image has support $[-2n; 2n]$). Formally, it can uniquely and equivalently be represented by $n$ subband signal functions $s_k$ (with $k = 0, 1, \ldots, n-1$), where the Fourier image of $s_k$ is contained in $[-2k-2; -2k] \cup [2k; 2k+2]$. Then

- $s(t) = \sum_{k=0}^{n-1} s_k(t)$

- Each $s_k$ has band-width 2, thus the corresponding signal can be perfectly reconstructed from samples taken at $1/2$ intervals.

- $s$ has band-width $2n$, thus it can be perfectly reconstructed from samples taken at $1/2n$ intervals.

Formally, each subband signal can be computed as an ideal band-pass filter applied to the input, i.e.

$$s_k(t) = s(t) * f_k(t)$$

where the impulse response of the $k$th (ideal) band-pass filter $f_k$ can be written as:

$$f_n(t) = \mathrm{sinc}(t) \cdot \cos((2n+1)t\pi)$$

(Note that $f_0(t) = \frac{1}{2}\mathrm{sinc}\left(\frac{1}{2}t\right)$.)  sinc does not have finite support in the time domain, so neither have the band-pass filters with impulse response $f_k$. Numerically the computations can therefore not be performed in the form written above – the ideal band-pass filters are approximated using the "windowing" method mentioned in section 1.3.2.2, with coefficients for a windowed sinc approximation specified in the standard[24].

MPEG audio statically partitions the frequency band into 32 equally sized subbands, with each subband sampled at $1/32$th of the critical sampling frequency of the original signal. Figure 1.12 shows the first 3 subband filters. Samples for all subbands are taken at the same point in time, thus they can be interpreted as 32-tuple "vector" samples. These subband samples can be stored in one of three possible formats (called *layers*), corresponding to increasing sophistication (and compression ratio):

---

[24]The standard does not specify the coefficients directly, but instead gives a method to incrementally compute the subband samples. From this procedure the underlying 512 coefficients can however be derived easily.

## 1.4. COMPRESSED REPRESENTATIONS

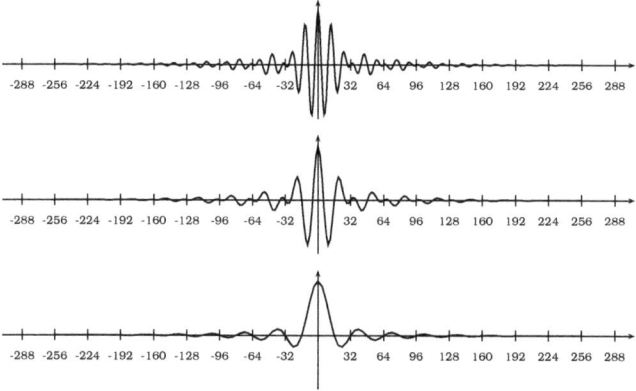

*The graphs above depict the analysis and synthesis filter kernels for the three lowest subbands (with increasing frequency from bottom to top) used for MPEG audio coding. The filter kernels have a support of 512 samples each, as a result each subband sample affects up to 256 samples of the reconstructed signal before and after the temporal position the subband sample was taken.*

Figure 1.12: MPEG audio analysis and synthesis filters

- Layer I takes 12 samples per subband to form a group. Each group of subband samples is coded separately using a scale factor and coefficients for the samples to adapt quantizing scale and dynamic range per subband. A block containing coded $12 \times 32$ samples for all 32 subbands forms a basic audio frame.

- Layer II is a straightforward improvement of Layer I coding: Subband samples are coded in 3 groups of 12 samples each. Layer II encodes one to three scale factors as well as the 36 coefficients per subband (thus already saving space if all three groups can be coded using the same scale factor), but using a more efficient (variable-length) representation than layer I. A block containing coded $36 \times 32$ samples for all 32 subbands forms a basic audio frame.

- Layer III does not directly store subband sample values. Instead, an MDCT (modified discrete cosine transform) is applied to blocks of 36 or 12 samples each, with $50\%$ overlap between blocks. In principle this doubles the number of coefficients, however half of the coefficients can safely be discarded as the samples in the overlapping area of two blocks can be reconstructed from the remaining MDCT coefficients (through time-domain aliasing cancellation [33]). MPEG layer III audio stores these MDCT coefficients, grouped into blocks of 18 coefficients and coded using variable-length codes. A block containing $18 \times 32$ coded MDCT coefficients for all 32 subbands forms a ba-

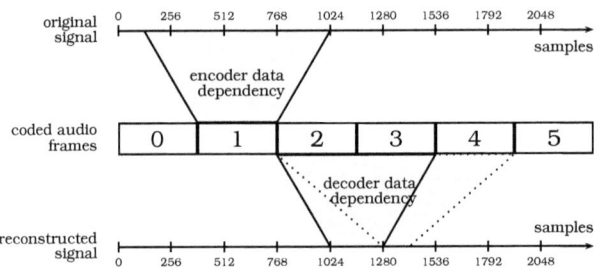

This figure illustrates the data dependencies introduced by the analysis and synthesis filter kernels of 512 samples width – each audio frame contains the temporal equivalent of 384 samples of the coded signal, but due to the "spread" of the filters, frame 1 depends on the values of samples 128 through 1024. A similar situation occurs for reconstruction: Samples 1024 through 1280 depend on data contained in frames 2 and 3, while samples 1280 through 1408 additionally also depend on frame 4.

Figure 1.13: MPEG audio layer I data dependency cones

sic audio frame. One peculiarity of the layer III format is the concept of a "bit reservoir": The bits comprising the coefficients may be spread over multiple frames, with forward/backward pointers stored per frame to reference the required data. This allows to "smooth out" bitrate variations between frames[25].

In all three layers the "partial" signals reconstructed from each block overlap if transformed back into a signal sampled at the critical sampling frequency (cf. figure 3.10 on page 104). Layer I and II frames are only "self-contained" in the sense that they allows reconstruction of the corresponding subband samples. Due to the "spread" of the filter kernels, reconstruction of a single sample may require as much as three consecutive audio frames (cf. figure 1.13). Layer III requires both the preceding as well as the following frame to undo the aliasing introduced by the MDCT just to reconstruct the subband samples – this leads to even wider data dependencies at the decoding stage (cf. figure 1.14).

MPEG audio achieves compression mainly by exploiting the spectral redundancy (i.e. "sparsity") of the frequency spectrum occupied by typical acoustic signals). The decomposition into subbands allows psycho-acoustic models to be applied that can identify signal components that may be suppressed because they are masked by other signal components (and are thus inaudible, cf. section 1.1.1.1). Note that the coding used in layers I and II potentially suffers from "blocking" artifacts as there is no "a priori guarantee" that the subband signals are smooth across block boundaries (see the discussion surrounding figure

---

[25]Essentially, this provides a mechanism for short-term varying bitrate with long-term constant average bitrate.

## 1.4. COMPRESSED REPRESENTATIONS

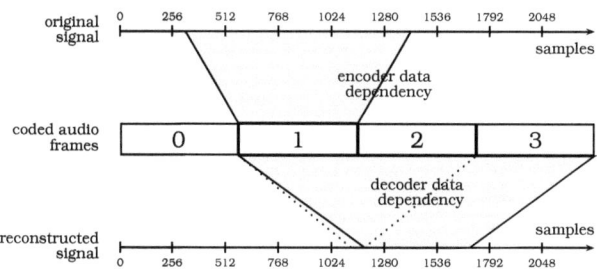

*This figure illustrates the data dependencies introduced by the analysis and synthesis filter kernels of 512 samples width as well as the overlapping for time domain aliasing cancellation after inverse MDCT. Reconstruction of samples 1632 through 1696 can be performed using the data from frames 2 and 3, but most of the samples depend on data from three adjacent audio frames for reconstruction (like samples 1696 through 2208).*

Figure 1.14: MPEG audio layer III data dependency cones

1.11), thus the signal must not be distorted too heavily to avoid discontinuities. The partial overlapping of subband samples introduced through the MDCT in layer III (and ensuing windowing) guarantees smooth transitions (individually in each subband signal, and logically also in their composition) even in the presence of distortions and thus also contributes to reducing the required number of bits for representation.

#### 1.4.1.4 Ogg Vorbis

Development of Ogg Vorbis began in the late '90s with the goal of producing a high-quality, freely implementable audio coding system. This was partly motivated by short-comings in the then-state-of-the-art MPEG Audio layer 3 compression, partly by the (perceived) legal restrictions on mp3. Despite lacking ratification by a recognized standards body it has found widespread support due to free availability of the specification [68].

While MPEG audio layer III uses a two-stage approach for frequency analysis of the input audio signal (subband transform followed by MDCT), Ogg Vorbis collapses the frequency analysis into a single MDCT[26]. This approach avoids some artifacts of the subband filter bank (signal aliasing in two subbands, difficulties in correlating higher spectral modes of musical sounds) and allows better global analysis of the signal spectrum.

Ogg vorbis groups MDCT coefficients obtained from one block transformation together to form an audio frame (as in MPEG audio layer III, blocks must partially

---

[26]In practice, a FFT is computed first as this representation is more amenable to spectral analysis. From the FFT coefficients the MDCT coefficients can be computed trivially.

overlap to perform time domain aliasing cancellation after inverse MDCT). The coefficients themselves are coded using a variety of coding tools offered by the format – first a rough approximation to the spectral curve (called the "floor" curve) is calculated, and parameters for this interpolant are coded. The sparse residual is then encoded through multi-pass vector quantization.

The Ogg Vorbis format allows considerable flexibility in representing the data and allows many parameters to be chosen by the encoder, including MDCT block sizes as well as codebooks for Huffman coding and vector quantization (in fact multiple parameter sets may be created and chosen individually for each new frame). These parameters are required for interpretation of frame data and must be available to the decoder before any processing can be performed. It should however be noted that each frame is self-contained and can be processed individually otherwise.

Despite the flexibility offered by the format, the decoding process itself is in fact not any more computationally expensive than MPEG audio layer III. The uniform use of the MDCT for spectral analysis also results in a considerably less complex data dependency chain for decoded samples: Each sample depends on data contained in exactly two adjacent frames.

**Summary**

The above discussion illustrates the principles behind modern compressed audio representations. While an exhaustive discussion would by far exceed the scope of this work, the vast majority of formats can be classified as structurally similar to one of the examples given above. For later purposes the following general structure should be observed:

- The compressed representation consists of
  - a bitstring $D$ (the *initialization data*), possibly of zero length
  - a sequence $F_0$, $F_1$, $F_2$, ... of bitstrings of "reasonably" limited upper length (the coded *frames*).
- The $n$th sample can be recovered from some tuple $(D, F_{i_n}, F_{i_n+1}, \ldots, F_{i_n+k-1})$ with fixed and "small" $k$.[27]
- Sample numbers monotonically map to frame numbers: $i_n \leq i_{n+1}$.

**Definition 13** *A (compressed) representation of a PCM audio signal is said to be* temporally local *if it satisfies the above constraints.*

It is easily verified that all example formats given in the preceding sections satisfy this requirement. While representations that violate the above restrictions

---

[27] The frame numbers $i_n$, $i_n + 1$, ..., $i_n + k - 1$ can be said to be the *decoding dependencies* of the $n$th sample.

## 1.4. COMPRESSED REPRESENTATIONS

can conceivably be constructed, the author believes these to be of little practical relevance. In particular, he is unaware of any representation in actual use that *cannot* be reduced to a combination of temporally local representations and the operators given in section 1.2.1.

### 1.4.2 Image Compression

Typical images found in multimedia applications have considerable *spatial redundancy*, i.e. samples corresponding to different but "close" points in an image are correlated (highly likely even similar). The common cause is that close points in the image are likely to correspond to different points on the same physical object, thus they probably differ only slightly in brightness and color tone due to different illumination and viewing angles.

The following sections will give a brief overview of common compressed image representations.

#### 1.4.2.1 GIF

GIF (short for *Graphics Interchange Format*) owes its popularity to Compuserve which introduced the format as graphics format for its online service in 1987. Its specification has subsequently been taken over by the World Wide Web consortium [14].

The format stores rastered images in an (unspecified) RGB color model (in absence of other information sRGB is usually assumed by most applications). Every GIF image consists of one or more sub-images, and each sub-image covers a rectangular area of the main GIF image. The sub-images are layered on top of each other in order to produce the final image[28]. GIF stores one or more *color tables*; each table may have up to 256 entries, and each entry is either an RGB color with $\alpha = 1$ (full opacity) or black with $\alpha = 0$ (full transparency). Color tables are stored uncompressed and require 3 bytes of storage space per entry: red, green and blue channel are stored with 8 bits of precision.

Every sub-image is conceptually a rectangular array of integers in the range 0 through 255 (and references a color table to supply an interpretation of each index value). These values are linearized in left-to-right and bottom-to-top order, and the resulting sequence is compressed using LZW.

GIF utilizes both limited color range (through the use of lookup tables) and spatial redundancy (the LZW compression represents recurring sequences of symbols through shorter bitstrings). However, this simple approach is largely ineffective in reducing redundancy if recurring patterns are merely "similar", instead of exactly identical: smooth color transitions (as they occur in many nat-

---

[28]A never formally specified (but accepted as a de-facto standard) feature allows simple animations to be stored as GIF: sub-images are not shown immediately, but after a delay time stored for every individual sub-image.

ural pictures) are particularly problematic, as 1) they require many color table entries and 2) produce many similar though not identical patterns.

#### 1.4.2.2 PNG

Development of PNG (short for *portable network graphics*) was incepted in 1994 and was subsequently recognized as a standard by several standards setting bodies (1997 IETF, 2003 ISO and W3C [15]).

Like GIF, the format stores rastered images in an (unspecified) RGB color model (with sRGB assumed usually). The image may either be stored in a color index mode (similar to GIF sub-images), or explicit RGB values for every pixel. Precision of color channels may vary from 1 to 16 bits; additionally an individual $\alpha$ value may be stored for every pixel.

PNG features improved redundancy elimination with respect to GIF by allowing to store not absolute but differential pixel values: the difference is taken with respect to a "predictor value" derived from a combination of the adjacent left, upper and upper-left pixel. The resulting values are then represented using a *deflate* encoding.

For images that are well-suited for GIF, PNG achieves at least comparable encoding, however due to predictive difference coding PNG is considerably more efficient at handling gradients and similar (instead of merely identical) patterns.

#### 1.4.2.3 JPEG

JPEG (Joint Picture Export Group) is an image format developed cooperatively between ITU [31], IEC and ISO; it has received formal acceptance as a standard by all three bodies.

The format stores rastered images with up to 3 color channels per pixel – normally $Y'P_bP_r$ or similar color models are used to decorrelate luminance from color tone, as in nature color tone is relatively uniform while luminance varies more widely. JPEG allows the stored color channels to be of different resolution – this to store reduced resolution versions of the less variant chroma channels.

The individual color channels are partitioned into blocks of $8 \times 8$ values, and a DCT (discrete cosine transform) is applied to each block. This transformation decorrelates the "average" value within a block (top left coefficient) as well as low to high frequency variations within the block (all other coefficients). The blocks are traversed from top to bottom and left to right. The top left coefficient of each block is represented differentially with respect to the previous block (on the assumption that adjacent blocks likely have similar average color), all coefficients in the block are quantized, serialized and represented using variable length codes (canonical Huffman codes). JPEG does not specify quantization and variable length coding tables but allows them to be embedded into the encoded image itself.

## 1.4. COMPRESSED REPRESENTATIONS

Compared to PNG and GIF, the JPEG format employs several transformations that on the one hand provide considerably better decorrelation (and thus better identify redundant information), but that on the other hand make the format unsuitable for pixel-exact representation of given images. In particular, the DCT is very effective in decorrelating smooth color transitions (i.e. it reduces the number of values required to represent the block's content to few non-zero values), but its main purpose in JPEG is to identify and filter out non-visible detail information that can safely be discarded. Therefore, the DCT is typically approximated using fixed-point arithmetic, and the "pseudo DCT" implementations in practical JPEG implementations are *not* invertible due to rounding errors (although DCT is mathematically invertible).

### 1.4.3 Compressed video

Time- and space-discretized video can be interpreted as a sequence of individual images, the individual images themselves can then be treated as described in section 1.4.2. This approach has the advantage that each individual frame of the video sequence can be stored and processed individually, thus it is used most often where access patterns require easy retrieval of individual pictures (i.e. video editing applications). As a disadvantage however this approach does not exploit the temporal redundancy present in typical videos, i.e. that temporally "close" pictures have similar features.

Several compressed video formats do therefore not represent each picture individually, but instead reuse features that have been encoded once for multiple pictures. Pictures that are coded "independently" are referred to as *intra* frames (I-frames), while pictures that reuse features from other images ("reference frames") are called *non-intra* or *predicted* frames. The most commonly taken approach for non-intra frames is to encode only the difference between pictures that contain the same or sufficiently similar features. Any compression schemes where individual pixels cannot be reconstructed from a single frame alone will be referred to as *temporal compression*. For later reference, the reader should take note that particular issues are introduced by formats that may use both temporally preceding as well as temporally succeeding images as "reference frames": Here, frames must be decoded in an order that differs from the temporal order in which the images are to be displayed, hence these formats will often be referred to as "out-of-order temporal compression". Two such examples will be given in sections 1.4.3.2 and 1.4.3.3.

The following sections will present several compressed video representations in widespread use, in order of increasing complexity.

#### 1.4.3.1 Motion JPEG

The Motion JPEG format stores video as a sequence of images; each image is individually stored using the JPEG format as discussed in 1.4.2.3. Rather than a formally defined standard, it is a convention used for interchange by many video tools. A description of the format can be found in [13].

Individual frames of the video sequence may either be stored as a single JPEG, or two JPEGs of half the vertical resolution each, representing even/odd fields of the video frame, respectively. This peculiar format is better suited for interlaced video data than a single JPEG image: the systematic difference between even/odd scanlines (that are usually captured at different points in time) would lead to vertical high frequency components in the corresponding DCT blocks: These components are usually most heavily distorted through the quantization process (leading to unpredictable "blurring" between scanlines), and at the same time reduce the number of bits available for representing other image detail that does not stem from recording artifacts.

Two variants, known as MJPEG-A and MJPEG-B, are commonly used. They do not differ conceptually, but have slightly different bitstream representations: while MJPEG-A retains the JPEG bitstream format (and thus embeds conformant JPEG images), MJPEG-B stores a slightly modified bitstream that identifies "sections" in the JPEG bitstream through an index table (instead of marker bits in the stream, as ISO JPEG does). The difference is only syntactic and transformation between the two variants is trivially achieved by adding/removing appropriate marker bits.

Unlike many other video formats, Motion JPEG has not received normative standardization; while "Apple QuickTime File Format" is widely regarded as authoritative, several implementations deviate from the format described therein. Nevertheless, the format has remained popular as support for Motion JPEG is basically "free" if JPEG still image (de)compression is available already.

### 1.4.3.2 MPEG-1/MPEG-2 Video

MPEG-1 video [30] is a compressed video format developed in the late '80s and standardized in 1991. It represents time- and space-discretized video as a sequence of individual rastered images represented in the $Y'C_rC_b$ color model. It employs a block discrete cosine transform for decorrelation similar to JPEG as described in section 1.4.2.3 to for decorrelation[29]. The syntax of the MPEG-1 format is quite flexible and allows a wide range of image sizes[30] and varying reproduction fidelity through the choice of quantization levels to be used for the sequence of images.

Images may either be coded *absolute* (without reference to temporally surrounding images) or *relative* (reusing features from temporally surrounding images), in particular MPEG-1 offers the following representation alternatives:

- I-frame ("Intra frame"): The image is encoded stand-alone.

---

[29]The similarity is however conceptual only: vast differences in the bitstream syntax, color model and subsampling strategies make translation between the formats not quite trivial.

[30]Syntactically, MPEG-1 is limited to $1024 \times 1024$ pixels per image.

## 1.4. COMPRESSED REPRESENTATIONS

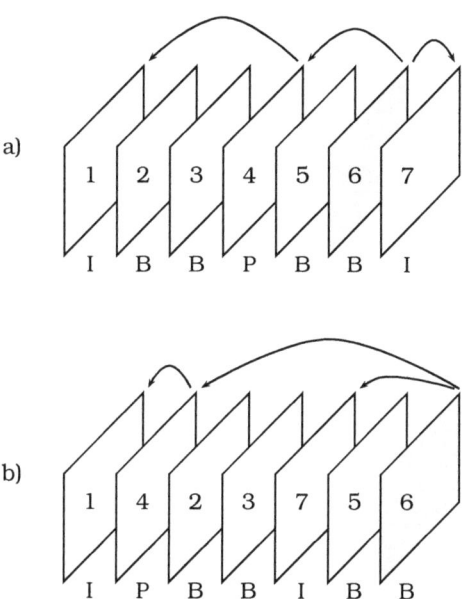

The figures depict images represented as I-, P- and B-frames. a) shows the images in temporal display order, while b) shows the same frames reordered such that all decode-dependent images follow their dependencies. The arrows depict the decoding dependencies of the sixth frame: It is coded relative to the fourth and seventh frames, but due to the fourth being a P-frame also recursively depends on frame number 1.

Figure 1.15: Frame ordering in MPEG-1/2 sequences

- P-frame ("Predicted frame"): The image is encoded *relative* to the temporally preceding I- or P-frame.

- B-frame ("Bidirectionally predicted frame"): The image is encoded *relative* to the temporally preceding I- or P-frame and the temporally succeeding I- or P-frame.

P- and B-frames can reuse features from reference frames by only encoding the difference of 16x16 blocks with respect to a selected 16x16 block of the reference image(s): The selected reference block does not have to be at the same spatial location; this is referred to as *motion compensation* and is achieved by encoding a displacement vector. The difference to the reference block may of course be zero, in which case parts of the reference image(s) are reused unaltered (e.g. static background, unchanging but moving objects).

# 48   CHAPTER 1. MULTIMEDIA REPRESENTATION AND PROCESSING

MPEG-1 video defines a bitstream format that contains a sequence of pictures coded using the coding options above. B-frames reference one temporally succeeding and one temporally preceding frame, so the temporal order in which frames have to be *decoded* does not necessarily match the order in which frames should be *shown*. The bitstream format stores frames in *decoding* order (cf. figure 1.15).

MPEG-1 can make use of the temporal redundancy by coding significant portions of a video as P- or B-frames[31]. Extensive use of non-intra frames introduces however two other problems:

- **random access limitations**: Extracting a specific image from a video sequence requires that all reference images are extracted as well (cf. figure 1.15). This in turn requires the ability to *identify* the dependency chain of an image.

- **error propagation**: P-frames may reuse parts of the previous P-frame with slight alterations, as a result decoding errors may propagate and accumulate in long chains of P-frames. This is especially problematic for MPEG as the transformations occuring during encoding and decoding (most notably DCT and inverse DCT) can – due to their complexity – in practice not be computed with infinite precision (and the standard does neither specify rounding directions nor error bounds).

The first problem is addressed through the specific format of the bitstream itself: Marker bitstrings are used to signify the beginning of a frame; applications can seek forwards and backwards from an arbitrary point in the bitstream looking out for these markers. Given the position of the desired image, applications can therefore simply "backtrack" to the preceding P- and I-frame(s). Additionally, the bitstream should contain GOP (*Group of Pictures*) headers that carry information about the point in time the current temporal position within the bitstream corresponds to, however this information is unfortunately not particularly reliable in practice[32].

In 1994 MPEG-2 (formally known as ISO/IEC 13818) was introduced as successor and refinement of MPEG-1. The main area of improvement is better support for interlaced video through various measures:

- Where MPEG-1 treats images as a "whole" like JPEG, MPEG-2 can (optionally) treat fields comprised of even/odd scanlines separately and thus

---

[31]The use of B-frames is however optional in the *constrained parameters* profile, and the use of both B- and P-frames is entirely optional otherwise.

[32]Note that a GOP is a *synchronization* and not a *decoding* concept (since GOPs are typically not closed and therefore may contain references to other images outside the GOP). Except for random access, GOP headers contain no information that could not be obtained by keeping counts of the number of frames processed already, decompressors *can* therefore safely disregard GOP headers altogether. This in turn has led to many encoders that do not bother to fill in GOP headers properly, so in practice a decoder *must* disregard GOP headers or otherwise be unable to process many streams.

## 1.4. COMPRESSED REPRESENTATIONS

support interlaced video without the issues already explained in section 1.4.3.1.

- Vertical subsampling of chroma values is optional in MPEG-2 (mandatory in MPEG-1). This means that even/odd scanlines can have different chroma sample values and therefore mainly also improves interlaced video.

- MPEG-2 supports motion compensation modes that "snap" to even/odd scanline boundaries of reference picture to make sure that even/odd fields of an image reference even/odd fields of the reference frame

MPEG-2 uses slightly altered codebooks for entropy-coding of DCT coefficients, motion vectors and macroblock modes; while these changes mainly account for the fact that MPEG-2 was targeted at higher image quality than MPEG-1 (resulting in altered ranges of DCT coefficients and thus different symbol statistics), the difference is rather minuscule in practice – except for interlaced images both formats are virtually identical in terms of coding efficiency. The most complete MPEG video reference is given in [32].

### 1.4.3.3  H.264 (MPEG-4 AVC)

Development of H.264 was incepted to provide a coding standard that further improves on MPEG-2 video in terms of coding efficiency, while still retaining a relatively simple decoder model. The format was standardized as MPEG-4 Part 10 in 2003 [29] but is also described in secondary literature such as [53].

Like MPEG-1/MPEG-2, H.264 makes use of predictive coding to reduce redundancy, but it features a number of novel ideas with respect to predecessor standards. Most of the coding efficiency improvements result from aggressive use of predictive coding at many levels. This includes:

- *spatial prediction*: While MPEG-1 and MPEG-2 only use the DC coefficient of the previous block as predictor for the current, H.264 uses all pixels adjacent to the current block.

- *temporal prediction*: Whereas MPEG-1 and MPEG-2 allow a frame to reuse features from up to two temporally close frames, H.264 puts no upper limit on the number of reference frames and their temporal distance. This is achieved in the bitstream format by storing frames in decoding order – each frame may contain a marker whether it must be retained for future reference by the decoder (possibly displacing a previously retained image) or can be discarded after presentation.

Predictive coding is however very sensitive to foward error propagation which must therefore be very tightly controlled. H.264 addresses this problem in a very radical way in that the image that must be reconstructed from a given compressed representation is unambiguously defined by the standard with no leeway for rounding or other errors. This results in a number of very radical changes of the algorithms used by decorrelation transformations – for example, where MPEG-1 and MPEG-2 use an $8 \times 8$ block discrete cosine transform, H.264 uses a much simpler $4 \times 4$ block transformation that are based on the matrices:

$$H_{forward} = \begin{pmatrix} 1 & 1 & 1 & 1 \\ 2 & 1 & -1 & -2 \\ 1 & -1 & -1 & 1 \\ 1 & -2 & 2 & -1 \end{pmatrix} H_{inverse} = \begin{pmatrix} 1 & 1 & 1 & 1/2 \\ 1 & 1/2 & -1 & -1 \\ 1 & -1/2 & -1 & 1 \\ 1 & -1 & 1 & -1/2 \end{pmatrix}$$

that are applied on each row and column vector of the block. Unlike the DCT, the matrices are orthogonal but not orthonormal, as a result $H_{inverse} \cdot H_{forward}$ reproduces the orignal values only up to a constant factor which must be taken into account for during encoding. The important property of this transformation is that it can be realized using just bit shift operations and additions (unlike the DCT which requires multiplication with irrational numbers) and is required to be implemented bit-exact.

Other improvements are achieved by more precise identification of object contours for motion compensation (up to single pixel precision, in contrast to $16 \times 16$ pixel precision offered by MPEG-1/MPEG-2). Additionally, H.264 also builds longer chains of temporally predicted images as prediction errors cannot accumulate.

**Summary**

The above discussion illustrates the principles of modern compressed video representations. While an exhaustive discussion would by far exceed the scope of this work, the vast majority of formats can be classified as conceptually similar to one of the examples given above. For later purposes the following general structure should be observed:

- The compressed representation consists of
    - a bitstring $D$ (the *initialization data*), possibly of zero length
    - a sequence $F_0$, $F_1$, $F_2$, ... of bitstrings of "reasonably" limited upper length (the coded *frames*).
- The $n$th $I_n$ image can be recovered from some tuple $(D, F_n, I_{i_{n,1}}, I_{i_{n,2}}, \ldots, I_{i_{n,k}})$ with some fixed and "small" $k$ and $i_{n,k} < n$.[33]
- The $k$th image to be displayed is some image $I_n$ with $n < k + d$ for some fixed and "small" $d$

**Definition 14** *A (compressed) representation of a time-discrete video is said to be temporally local if it satisfies the above constraints.*

---

[33]The frame numbers $i_{n,1}$, $i_{n,2}$, ..., $i_{n,k}$ can be said to be the *decoding dependencies* of image number $n$.

## 1.4. COMPRESSED REPRESENTATIONS

It is easily verified that all examples given above satisfy this definition. While representations that violate these restrictions can conceivably be constructed, the author believes these to be of little practical relevance. In particular, he is unaware of any representation in actual use that *cannot* be reduced to a combination of temporally local representations and the operators given in section 1.2.2.

# Chapter 2
# Related work

This chapter discusses existing media processing infrastructure work. The emphasis in the first half of this chapter is on "complete" media processing frameworks and their architecture. They serve as "role models" for the architecture developed in chapter 3, as well as setting the benchmark to compare this work with. So as not to exceed the scope of this work, these frameworks will however not be discussed in their entirety – instead the focus is on a select number of key aspects that highlight crucial design decisions and their consequences and thus prepares the reader to understand the design presented in chapter 3.

The second half will discuss the target environment of the media framework in chapter 3. This includes existing system services for various low-level media processing tasks such as video and audio capture and playback. It also includes a discussion of existing toolkits that offer services for higher-level media processing such as compression and editing.

## 2.1 Media processing frameworks

This section will investigate existing frameworks that facilitate processing of time-based media. Other than this subject matter, the different frameworks have few commonalities and mainly reflect their respective developers' thinking of media processing. The purpose of this section is to highlight both commonalities and differences in the approaches.

At the functional level, the different frameworks provide support for (at least some of) the following services:

- *Data representation*: Representation of media data in multiple formats as well as transformation between different representation alternatives (e.g. transformation into a compressed representation before storage, decompression before processing)

- *Capture*: Acquisition of media through physical devices (e.g. cameras, microphones) including the task of determining the timing relationship and/or synchronization of the data capture devices involved

- *Playback*: Reproduction of media through physical devices (e.g. displays, speakers) including the task of determining the timing relationship and/or synchronization of the playback devices involved

- *Persistence*: Storage and retrieval of media data (e.g. to the file system) including the task of (de)multiplexing related media

- *Compositing and Processing*: Creating new media by compositing or otherwise processing existing media (e.g. mixing, filtering)

However, a "mere collection" of these functional components alone does not make a "framework": The term *framework* implies several architectural (in addition to the functional) characteristics:

- *Abstraction*: Different entities providing the same class of functional service are accessible through a common interface. For example, ADC converters for audio capture may be connected to the host system through various physical connectors (e.g. PCI bus or USB) and require different access mechanisms at the driver software level – at the level of the media framework they are expected to be accessible through the same interface[1].

- *Modularity*: Different entities providing different classes of functional services can be freely combined by applications to process media data as they wish. For example, media captured through *any* of the acquisition devices can be further processed through *any* means the framework offers, and played back through *any* reproduction device[2]. While this requirement appears trivial at first, it can become technically challenging to meet when considering this requirement against *efficiency* – "combined" processing of several steps is generally more efficient than the modular composition of the indivdual sub-steps.

- *Extensibility*: New entities providing a suitable abstract interface defined for one of the functional services provided by the framework can be added, and can be used transparently as if it were a built-in service of the media framework. Examples include accessors to different media storage containers, or different media (de)compressors.

This essentially suggests a *component*-based architecture, where the frameworks defines the component interfaces as well as their interactions and provides methods for the management of component classes and instances. This includes:

---

[1] Note that different layers in the software stack might provide the abstraction (e.g. at the driver interface, operating-system provided wrapper libraries, or an abstraction layer within the media framework) – thus this requirement captures only the user's expectation and is not necessarily an implementation requirement for the media framework writer.

[2] though possibly subject to *technical* restrictions such as communication channel capacities and available processing power – the key point is however that no *architectural* restrictions remain – provided the technical restrictions could be overcome.

## 2.1. MEDIA PROCESSING FRAMEWORKS

- *Registration*: Keeping track of available components and allowing dynamic registration for extensibility
- *Lookup*: Methods to select component classes according to various criteria; this may include the capability to browse the set of elegible components (e.g. to select a desired compressor)
- *Instantiation*: Creating instance objects of component classes to access the functional service

These functional aspects can be found in all frameworks discussed below, but the exact realizations in the respective architectures differ considerably. The following will focus on illustrating these different design choices.

### 2.1.1 QuickTime

QuickTime is a media processing framework developed by Apple Computers. Originally released in 1991 for Macintosh System 6 computers, it has been in continuous development until present. Changes introduced over time extend the framework with new capabilities (e.g. introduction of distinct decoding and display order for images, QuickTime 7 in 2005), or have adapted the framework to changes in the operating environment (e.g. deprecation of `FSSpec` for identification of files in favor of a more flexible URL-based approach, QuickTime 6 in 2002). Though core concepts have remained unchanged, the description given here is always based on the latest available version at the time of this writing (QuickTime 7) where behaviour deviates or interfaces have been deprecated with respect to older versions.

**Terminology**. QuickTime reflects concepts of object-oriented programming reasonably well[3], however it predates modern software engineering terminology, and most available QuickTime literature favors the "traditional" terminology. For easier comparison with the other architectures presented in this chapter, modern terminology will be prefered.

#### 2.1.1.1 Overview

QuickTime is conceived as a modular architecture with most of the functional services provided by replaceable *components* and *component instances*. Each component instance provides a specific interface defined by the QuickTime architecture. Component instance classes that derive from the same abstract interface can generally be used as substitutes for each other (component instance classes are not necessarily direct ancestors of their respective interfaces but may derive from intermediate component instance classes[4]).

---

[3] e.g. the opaque **Movie** data structure can be regarded as an object, with many QuickTime API functions taking the role of methods on this object.

[4] QuickTime provides a mechanism for component instances to "encapsulate" other component instances and selectively use or override the functionality of the enclosed instance. This

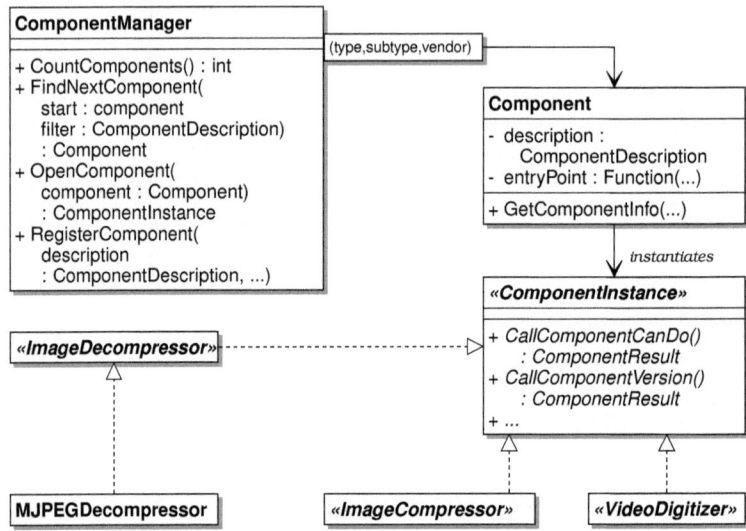

QuickTime defines abstract interfaces such as **ImageDecompressor** which are implemented by concrete classes such as **MJPEGDecompressor**. The singleton component manager provides methods to register and lookup components and to instantiate the concrete classes (via delegation to the underlying component) that provide e.g. mechanisms to decompress media data

Figure 2.1: QuickTime component manager

The components are organized by "type", "subtype" and "vendor" which each consist of a fixed 4-character string. The "type" of a component specifies the interface to be expected of the corresponding component instances (and thus their purpose for which they can be used in the context of media processing), a non-exhaustive list of different pre-defined component types is:

- `imco` (image compressors): Provides means to convert a rastered image (cf. section 1.3.4.1) into a different (compressed) representation.

- `imdc` (image decompressors): Provides means to convert a compressed image representation into a rastered image.

- `aenc` (audio compressors): Provides means to convert PCM audio samples (cf. section 1.3.2.1) into a different (compressed) representation.

---

allows for the functional equivalent of implementation inheritance to be realized between QuickTime components, however it should be noted that the technical realization of the dispatching mechanisms is very different from that commonly used in object-oriented languages. While deriving components from existing components is generally encouraged (e.g. [52] p. 11), it could not be determined if the pre-made components generally shipped with QuickTime follow this practice.

## 2.1. MEDIA PROCESSING FRAMEWORKS

- `vdig` (video digitizers): Can capture video data from an external source.
- `eat` (movie importers): Import foreign container formats as movies.
- `spit` (movie exporters): Export movie to foreign container formats.

A component's "subtype" and "vendor" are used to distinguish different components that provide functionally equivalent services through a common interface. By convention, components with common "subtype" can process data of the same underlying format (e.g. decompressors for different image formats) while different "vendor" strings merely distinguish different implementations of processors for the same format – however neither is this convention really relevant for all types of components, nor is it followed rigorously in practice.

Applications as well as the rest of the QuickTime architecture use the *component manager* to select and instantiate components. Conceptually, the component manager's role is to maintain the mapping from the namespace consisting of triples of the form (type, subtype, vendor) to the respective component classes, and thus supports the usual operations expected for the management of namespaces (registration, lookup, browsing). The namespace operations supported by the component manager are closely tied to the desired interpretation of the name tuples in the context used by QuickTime: generally components are requested using only the (type, subtype) tuple part of the name, with vendor left as "wild card" (configurable preferences can establish a "search order" among all matching components).

QuickTime supports dynamic loading of components compiled as loadable modules transparently – obviously, the components cannot be registered through the run-time call **RegisterComponent**. The component manager therefore supports a secondary registry in persistent storage, which is realized simply as a directory containing the dynamically loadable libraries and which is scanned by the QuickTime library during initialization. To avoid actually loading all libraries (which would incur significant overhead as it requires processing by the run-time linker for relocation and symbol resolution) each is accompanied by a "resource file" providing information such as the (type, subtype, vendor) and entry point into the loadable module.

### 2.1.1.2 Movies

The highest level of abstraction offered by QuickTime for time-based multimedia data is that of a *QuickTime movie*, represented in the architecture through the **Movie** interface. Conceptually, a QuickTime movie is an aggregation of multiple time-based media elements (called **Tracks**) with a defined temporal relationship (cf. figure 2.2). The **Tracks** in turn contain one **Media** object each, through which the actual media data is accessible. The separation of the **Track** and **Media** concepts reflects the separation between the temporal position within the movie and the media data itself (as well as its logical position within the data store). Conceptually, each track can contain arbitrary time-based media (the type of media

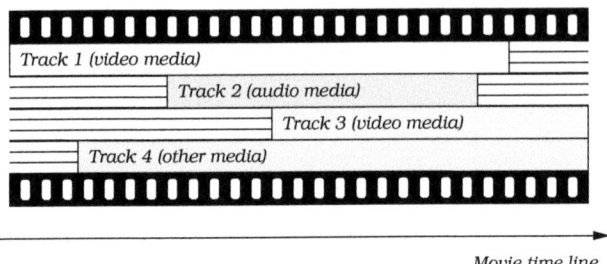

Figure 2.2: Conceptual view of a *QuickTime movie*

is identified through a four letter code as usual) – while video and audio are the most commonly used types of media, others can be defined and processed by creating appropriate data handler components.

QuickTime assumes that media is stored as a sequence of equidistant time-discrete samples (cf. sections 1.3.2.1, 1.3.5.1) organized into *frames* that can (largely) be processed individually. Each frame is assumed to be stored as an opaque sequence of bytes that can be expanded into a single rastered image or a fixed number of PCM audio samples using an appropriate interpretation of the data (the mechanism QuickTime uses to determine this interpretation will be discussed further below). Applications that so desire can directly access each individual frame through the **GetMediaSample/GetMediaSample2** calls on **Media** objects.

QuickTime provides some support for random-access to **Movies** using temporal prediction for video coding: Individual frames can be marked as *key frames* to indicate that no following frame has a data dependency on previous frames. This annotation is independent from the format the frames are coded in, so applications can obtain this information without understanding the format.

While the QuickTime API uses **Movie** to represent QuickTime movies, **Movie** objects also carry a considerable amount of state information (e.g. brightness adjustments for display, error states of previously executed operations, current playback position) that is not actually reflected in the underlying media data itself, but used for playback or other forms of access to the movie. The API thus blurs the distinction between *data* and *accessor objects* to the data. Generally, all accesses to the movie data must be multiplexed by the application through the corresponding **Movie** object (with possibly conflicting state information saved and restored) – QuickTime does not provide any mechanism to instantiate additional **Movie** objects as accessors to the same underlying conceptual QuickTime movie[5].

**Movie** objects are generally instantiated in one of two ways: Either as pure in-memory "scratch" movies, or from a file-based data store. While QuickTime

---

[5]Multiple **Movie** objects can be instantiated as accessors to a movie stored as a file, however the behaviour is undefined if write access is performed through one of these **Movie** objects.

## 2.1. MEDIA PROCESSING FRAMEWORKS

can read and write movie data into a variety of different file-based container formats, only the QuickTime file format is treated as "first-class citizen": Access to data in other formats is achieved through **MovieImport** and **MovieExport** components, however these component interfaces can only support either read-only or write-only access to alien formats. Thus, all of the editing functionality normally available through **Movie** objects is unavailable, so **Movie** objects provide a rather leaky abstraction.

### 2.1.1.3 Compressed media data

While **Movie** objects allow to manage, store and retrieve media data from and into container objects, they do not interpret the contained data. QuickTime does not provide abstractions (in the sense of an *abstract* or *interface* class) for the media elements it supports, such as *image*s. This means that while the media elements certainly exist as concepts, it is the application's responsibility to associate desired semantics with the data retrieved from **Movie** objects. QuickTime only supplies a **SampleDescription** that carries meta-information (e.g. picture width/height) as well as a four-letter code identifying the format. The sample description may also contain further media-format-specific data (e.g. initialization data for decompressors) that is required to interpret the data.

The four letter code contained in the **SampleDescription** is interpreted by QuickTime as the identifier to be used for looking up a required decompressor component using the component manager[6]. For QuickTime movie files this is the same code as stored in the on-disk representation of the movie. For other container formats (e.g. AVI) the **MovieImport** component is expected to perform the required translation between the code stored in file and the code expected by QuickTime (thus every **MovieImport** component must know about every possible translation for their respective format as QuickTime does not provide a registry for this type of translations).

It is up to the consumer of the data received from a movie to utilize the appropriate decompressor components when reading data out of a **Movie** (or, vice versa to instantiate the compressor components when writing). For image data, this means instantiating an **ImageDecompressor**(imdc) component, for audio data an **AudioConverter** (adec) component. These component types expect the data for individual frames to be fed in correct order, and will decompress them in sequence. If decoding dependencies exist between frames, the components are required to detect them and internally hold necessary state (e.g. old pictures used as reference frames for future pictures) – this process is transparent for the supplier of the compressed media data.

The decompressor API provides separate methods that initiate processing of a frame (**ImageCodecBeginBand**) and methods that finally output the decoded data (**ImageCodecDrawBand**) to a designated destination. Since QuickTime 7

---
[6]Normally, this code also identifies the *compressor* that can be used to compress the rastered image or PCM audio data into this particular format, however this symmetry is not universal.

this interface explicitly allows frames to be displayed in a different order than decompression was initiated. The decompressor components are ignorant of the temporal relationship of the frames processed, the correct point in time for playback is part of the **Movie/Track** structures containing the data. Applications processing media data must therefore feed data to the decompressors in the order they are contained in the **Media** structure, but schedule playback or other processing using the time information from the containing **Track** and **Movie**.

Compression of media data is performed in analogous fashion – rastered image or audio sample data is fed to the compressor instance which transforms it into a compressed representation. Image compressor components use flags to advertise the capability of their underlying format to support frame sequences that are temporally out-of-order, however in this it is case the compressor that decides on the ordering of frames and passes this information back to the caller. It is then the caller's responsibility to store the data into a container with appropriate timestamps that allow reconstruction of the original temporal sequence.

### 2.1.1.4 Compositing and processing

QuickTime generally allows the applications direct and easy access to the media data in its various forms of representations, i.e. compressed frames or uncompressed rastered images and PCM sample data. Any further processing of the data is largely out of the scope of QuickTime, and the application must generally use other means such as the Quartz graphics API to manipulate images. This is not necessarily to be seen as a conceptual "weakness" of QuickTime as it provides applications the flexibility to use othe tools of their choice to achieve the desired effects.

But QuickTime also provides a form of "compositing" support for video media through the concept of pluggable video effect components: These operate on one or more input images to produce a single output image and are thus a generic form of an image processing operator (cf. section 1.2.2). There are numerous predefined effects that correspond to different types of scene transition effects; many of these are realized as unary image processing that affect only of the images (by e.g. applying desired transformations and adding an alpha channel) which is then composited over the second image (see also section 2.2.2.1 for an example of this technique).

The video effects may be parameterized, and the parameters may be varied over time to provide desired temporal transition effects. Parameters may be controlled directly by the application, but QuickTime also supports storing these in a special track as part of a **Movie** data structure.

At a technical level, video effects are closely related to QuickTime decompressor components: They provide the same component interface and thus may be substituted wherever a decompressor would be valid. QuickTime also provides a simple mechanism to automatically "chain" multiple filters, so the last chained filter appears as a singular source of video data.

## 2.1. MEDIA PROCESSING FRAMEWORKS

### 2.1.1.5 High-level data handling

QuickTime provides further assistance in relieving application programmers from the low-level tasks of handling media data. For images, the *image compression manager* can take over most of the tasks described in the previous sections. Even one level higher, QuickTime provides the concept of **MediaHandler** (`mhlr`) components – they completely encapsulate all tasks required to retrieve data from a `Movie`, interpret it appropriately and further process it. **MediaHandler**s are implemented for specific types of elementary media (e.g. video or audio media), typical **MediaHandler**s know how to interact with the display and audio systems to facilitate playback.

QuickTime provides even higher-level services, such as user interface elements for the reproduction of video including VCR[7]-style playback controls (movie player). These should arguably however not be considered integral parts of the architecture as their role is to provide a "bridge" between two conceptually very different frameworks (multimedia processing and graphical user interfaces).

### 2.1.1.6 Capture

Capturing of image sequences is performed using the **VideoDigitizer** (`vdig`) component interface; for audio capture, QuickTime does not provide a component interface but requires the application to access the audio device using the low-level system APIs (e.g. CoreAudio).

The interface provides various methods to query the underlying driver's capabilities and requires the application to setup several parameters before capturing can start. Two sets of different API functions can then be used to obtain the frame data, depending on whether the **VideoDigitizer** provides compressed or uncompressd frame data.

For uncompressed data, the application may either use "continuous digitization"; in this case it sets up a single buffer (**VDSetPlayThruDestination**) and starts digitization (**VDSetPlayThruOnOff**) – it then becomes the application's responsibility to process the captured data in time before it gets overwritten. The second option is to use asynchronous single frame grabs – in this case the applications sets up (**VDSetupBuffers**) a number of buffers that can hold one frame each. It can then issue multiple **VDGrabOneFramAsync** calls that instruct the digitizer to start operation and fill the supplied buffers one by one. The application must use the **VDDone** call to check if any asynchronous frame grabs issued previously have completed. Afterwards the application may process the received data (and reissue a new frame grab operation).

For compressed data, the **VideoDigitizer** component only provides an interface equivalent to the asynchronous interface of uncompressed data capture: **VDCompressOneFrameAsync** issues a request to grab one compressed frame, **VDCompressDone** checks whether any issued grab request has completed. However, in this case the **VideoDigitizer** component "owns" the memory

---
[7]video camera recorder

allocated to hold the compressed frame data and must thus be returned through **VDReleaseCompressBuffer** to the digitizer.

Interestingly the API does not provide a method that blocks the calling thread until completion of an issued asynchronous operation (as would be common architecture practice using the *active object* and *future* patterns) – the only operation provided is to register a callback procedure (**VDSetDigitizerUserInterrupt**), forcing the programmer to either adopt an event-driven programming model or resort to periodic polling.

QuickTime also provides more high-level components such as the **Sequence-Grabber** (grab) interface that automates most of the tasks involved above and provides mechanisms to automatically transfer the data into a **Movie** data structure.

### 2.1.1.7 Discussion

The QuickTime architecture offers a set of building blocks that application developers can combine in flexible ways to realize media processing. The building blocks are structured in a logical way, and – perhaps more importantly – are generally useful on their own. The API itself represents concepts of object-oriented architectures reasonably well (due to its heritage it is however quite idiosyncratic since it does not follow established syntactic and architectural conventions for many common concepts). Nevertheless, the architecture exhibits some fundamental shortcomings that should be mentioned.

The individual component classes of the architecture are orthogonal at a functional, but not at a technical level. Taking for example the task of decoding a compressed image, one could realize the multiple steps typically required in different ways. It might be desirable to perform entropy or other symbol decoding on the host CPU while delegating computationally expensive transformations (e.g. discrete cosine transform) to a dedicated processing unit located at the graphics card. However, this delegation only makes sense if the resulting image is intended to be displayed by the graphics hardware later – if the image is to be further processed by the host CPU it is typically more efficient to have the CPU perform all required transformations itself as the required data round-trips across the system bus would by far exceed any potential performance gains. This requirement poses a problem for the simple image decompressor model featured by QuickTime: From the perspective of clean separation of concepts in the architecture it is undesirable that the decompressor has knowledge what the application intends to do with the resulting decompressed image.

QuickTime provides a mechanism that allows decompressors to short-circuit with the display system, thus breaking orthogonality of the components at the technical level[8]. While this approach addresses the performance problem in a special (albeit important) case, it immediately raises doubts about the approach

---

[8]Note that this is a direct result of the design of the exported component *interfaces*, it would thus require an incompatible interface change to address this issue cleanly.

## 2.1. MEDIA PROCESSING FRAMEWORKS

taken in the architecture and begs the question if a different approach might have addressed the problem in a cleaner and more general way. Similar (and worse) but more subtle interactions can be found in the audio subsystem.

The data model offered by **Movie** objects is not completely convincing either. Since modern video and audio coding techniques make extensive use of temporal prediction (cf. section 1.4.3.3), it is inevitable to deal with the complex issues brought forth by out-of-order decoding. However, QuickTime offers little assistance to application programmers as they have to either use the (relatively opaque) **MediaHandler** interface and divert their output, or deal with the full complexity themselves. Moreover, the simplistic approach to annotate data dependencies purely through the concept of *key frames* is problematic – in complex scenarios a sequence of images may have *no* key frames at all, even though each image has only finite and temporally close data references[9].

The QuickTime low-level API is centered around the idea of frame-level random access semantics – which is conceptually nice as it provides a great amount of detailed control over the media processing – yet QuickTime fails to consistently maintain this semantics as its level of abstraction is too low for this purpose.

### 2.1.2 DirectShow

DirectShow is a media processing framework developed by Microsoft. It was originally released under this name in 1997, but is based on older frameworks such as Video for Windows and ActiveMovie. Despite its origin as a pure media playback architecture, it is quite capable of expressing complex media processing and editing operations.

#### 2.1.2.1 Component model

DirectShow relies on the generic COM+ component object model that is already an integral part of the Windows run-time environment. As such, COM+ takes over all tasks of managing, browsing, querying and instantiating components – the "only" task left to DirectShow is to define the component interfaces accessible through the usual COM+ mechanisms.

DirectShow implicitly uses many services provided by COM+, for example applications can use the run-time type information facilities to query whether a particular component provides a desired interface (e.g. **IMediaSeeking** providing mechanisms to perform random seeks to desired points in time). However, discussion of COM+ is out of scope for this work, the reader is referred to the relevant literature [54] instead.

DirectShow relies on the COM+ system registry to associate "names" (GUIDs) to component classes. In most cases DirectShow does not perform name-based

---

[9]Consider an MPEG-2 video coded as IBIBIBI... in display order or IIBIBIB... in coding order – each B frame can be decoded using the surrounding I frames as reference, yet *none* of the I frames beside the first can actually be a key frame in the sense used by QuickTime. See also figure 1.15 on page 47.

# CHAPTER 2. RELATED WORK

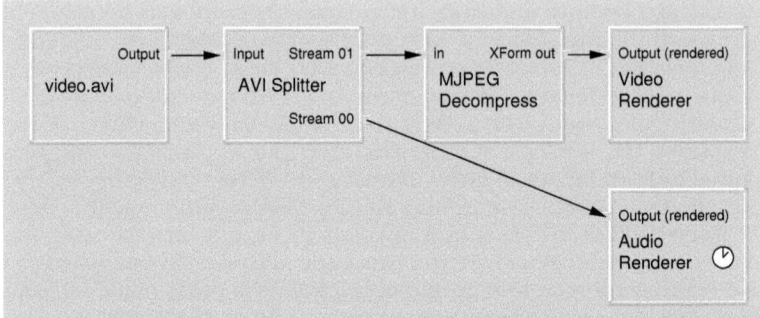

*The above graph represents the playback of an AVI file: The left-mode node (video.avi) reads raw octets from the underlying file, the AVI Splitter node is responsible for extracting desired tracks from the stored media. The demultiplexed video track needs to be decompressed (using the motion JPEG decompressor in this example), decompressed video and audio stream can finally be rendered. The Audio Renderer also provides the reference clock for this presentation.*

Figure 2.3: Example DirectShow filter graph

lookup, but uses more complex criteria such as input/output format constraints, semantic annotations of connector pins and filters to find component classes. For this, DirectShow maintains several other registries that map to the GUIDs of component classes. In addition to the constraint-based lookup DirectShow also relies on the system component enumeration mechanism where multiple functionally equivalent components can fill in a desired role (e.g. selecting video input devices).

#### 2.1.2.2 Processing concepts

Media processing in DirectShow is centered on building a *filter graph* that describes the flow of data (edges) and processing steps to be applied to the data (nodes, called *filters*). The filter graph itself is represented through an object that implements **IFilterGraph** interface which contains methods for adding, removing and locating filters (it is however recommended to use one of the derived interfaces like **IGraphBuilder** instead, see below). It is also responsible for synchronizing all media processing operations to a reference clock (represented through the **IReferenceClock** interface, usually one of the objects contained in the graph itself will take over this role) and provides other interfaces such as **IMediaControl** that allow to start, stop and pause processing. Figure 2.3 shows an example filter graph.

DirectShow defines the interfaces **IMediaSample** as a generic container for any data passed between two nodes in the filter graph. In addition to the actual media data, each **IMediaSample** may also transfer additional meta-data such as timestamps that help in interpreting the data. The timestamps also allow the

## 2.1. MEDIA PROCESSING FRAMEWORKS

filter graph object to schedule operations: Normally, the processing of the data progresses in "lock-step", but **IMediaSample** objects that do not carry timestamps are exempted from this rule.

The filter nodes themselves are represented through the **IBaseFilter** interface and contain the actual processing logic. At the functional level the individual filters can provide a variety of services in the context of media processing, for example:

- *Source filters* have only output pins and supply "initial" data into the graph; this can be synthesized data or any other data that needs to be inserted into the media processing from the outer world (e.g. data read from a file).

- *Sink filters* have only input pins and receive the end results of the filter chain. They may e.g. write compressed data into a file, or *Renderer filters* in particular may be used to playback media.

- *Mux/Demux filters* either multiplex multiple data streams from multiple input pins to one output pin, or vice versa.

- *Format conversion filters* transform given data into a different format. They are usually only introduced as auxiliary filters to satisfy input/output constraints of filters that are to be connected when completing the graph.

The architecture allows arbitrary other types of filters that perform application-specific transformations on the data.

Each filter possesses a number of input and output pins (implementing the **IPin** interface). The pins express their filter's input and output format constraints (an output pin may be connected to an input pin only when the sets of supported formats are not disjoint) as well as some semantic information for the filter graph manager (see below). They are also responsible for negotiating the format actually (if multiple are elegible) as well as the method of data transfer to be used between the connected nodes (e.g. push/pull using callbacks, or through interthread communication).

While applications can build their own filter graphs from scratch, it is generally recommended to use the filter graph manager through the **IGraphBuilder** or other more specialized interfaces – they provide high-level methods that build a complete graph from given "graph fragments". The filter graph manager will search the component registry for conversion and (de-)multiplexing filters such that it can form a valid graph.

It should be noted that each of the objects (filter graph, filters and pins) typically supports other interfaces – for example data source and sink filters may provide the **IMediaSeeking** interface and advertise their capability to jump to specific positions in time. The filter graph object in turn will be aware if the aggregate of all filter nodes is capable of seeking and may also expose this capability through the **IMediaSeeking** interface.

### 2.1.2.3 Data representation

DirectShow uses **AM_MEDIA_TYPE** objects to identify different media formats, both for matching of connectable **IPin**s as well as attached to annotate each **IMediaSample**. The architecture is not limited to a fixed set of different formats, but can be dynamically extended as well. As to what constitutes a distinct "format" is generally up to the implementor of filters consuming and producing the corresponding data. Typical examples are raw (unprocessed) byte-stream data, demultiplexed audio or video data, individual frames of a specific compressed format, or rastered images in a specific colour model.

Note that the data contained in an **IMediaSample** can usually not be interpreted outside of the context of the filter graph – while some formats may choose to represent e.g. self-contained images in a single **IMediaSample**, this is pretty much the implementor's choice. In general, issues such as reference frames or out-of-order decoding become an implementation detail of the filters which may have to perform requisite buffering. The architecture is centered on the concept of *streaming* media data through the filter graph that not too much semantics should in general be put into individual **IMediaSample** objects.

### 2.1.2.4 Capture

Capture of media data (video and audio alike) is performed by instantiating corresponding filter nodes as "sources" for the whole graph: These nodes do not have any input pins, but at least one output pin to represent the captured data.

The filter objects representing data capture are not instantiated through DirectShow; instead the application must use the system device enumeration service to locate available devices from which the application must choose. The devices in turn provide a standardized interface for instantiation of data capture filters that can be added to a filter graph.

### 2.1.2.5 Editing services

While the filter graph concept of DirectShow allows to express complex media processing operations, it is very inconvenient to manually construct such a graph – in particular, the filter graph only represents the "static" (time-independent) view of the processing chain: Effects that are to be activated at specific points in time must be followed by *delay* or temporal *multiplex* filters – the view of the graph thus shows all effect filters ever applied during the processing, regardless of the point in time they are supposed to be active.

A more suitable representation for media editing is through *time lines*: these provide a time-variant view of the individual media including the effects active at any given point in time (cf. figure 2.4). DirectShow editing services consists of several APIs that allow to manipulate time lines, serialize and deserialize time lines to/from storage, and convert time lines into a filter graph representation

## 2.1. MEDIA PROCESSING FRAMEWORKS

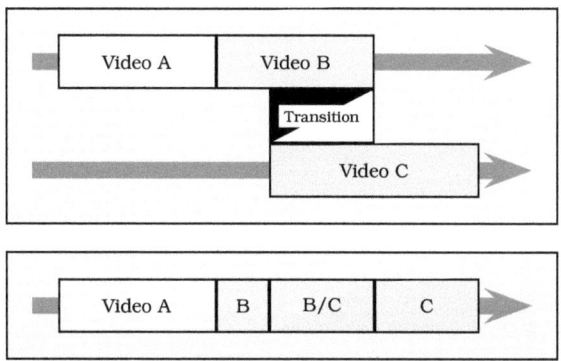

*The upper half shows a video consisting of two logical video time lines as well as a transition between the two: The first scene consists of video A, the second scene starts with video B but gets faded over smoothly using a transiation effect into the video C. The lower half shows the composited video. A filter graph for realizing this composition is shown in figure 2.5*

Figure 2.4: Time line view of a video editing operation

to actually perform the processing (cf. figure 2.5). The resulting filter graph can then be used to render the resulting multimedia presentation to the display for preview or into a file by instantiating corresponding render nodes to terminate the graph.

### 2.1.2.6 Discussion

DirectShow provides a comprehensive framework to express media processing through its very abstract and generic filter graph concept. The underlying COM+ model provides considerable flexibility in extending the architecture and adapt it to the specific processing requirements, however it introduces tremendous complexity for the component implementors.

The streaming media approach employed by DirectShow simplifies integration of modern coding techniques using both forward and backward prediction of frames as the processing nodes can (and actually *must*) transparently buffer and reorder frames (in contrast to QuickTime's approach where the temporal relationship is completely exposed). It should however be noted that it incurs significant semantic loss compared to QuickTime as the structure of the underlying media data becomes completely opaque – basically, any correlation between e.g. compressed input data and uncompressed output data of a decompressor filter is lost, as the application simply cannot assume that the filter is stateless. This also introduces new problems when frames must be dropped (e.g. due to corrupt/lost data or due to system overload) – the stateful nature of the filter nodes and lacking semantics of data flowing through the graph make resynchro-

# 68 CHAPTER 2. RELATED WORK

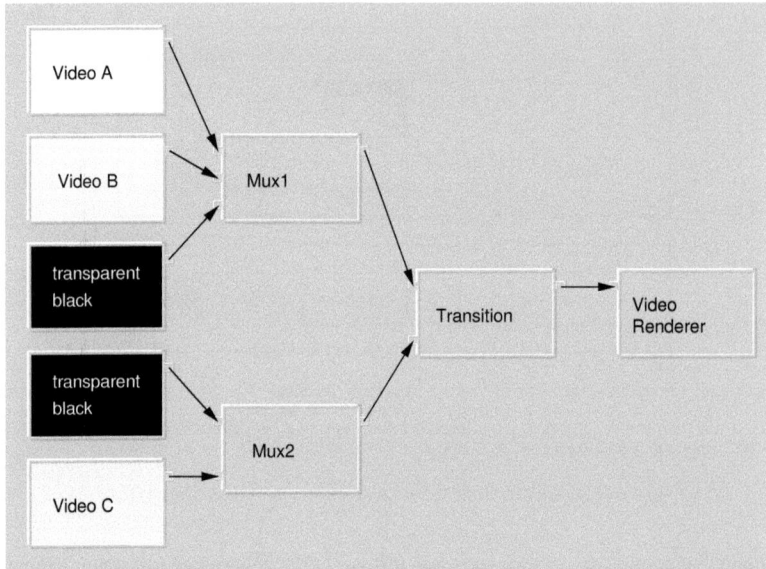

*The above graph represents the compositing operation shown in figure 2.4 as a filter graph: The nodes Mux1 and Mux2 perform a temporal multiplex of the input video scenes that correspond to the upper and lower time line of figure 2.4, substituting a transparent black image when no video has been defined for the corresponding point in the time line. The Transition node finally realizes the compositing of the two time lines, blending the images from video C smoothly over the images from video B at corresponding points in time.*

Figure 2.5: DirectShow filter graph realizing a compositing operation

nization with an interrupted data flow extremely difficult and in some cases even impossible.

DirectShow's streaming approach also provides a clean mechanism to conditionally delegate parts of the media processing operations to the graphics hardware (cf. section 2.1.1.7): Media data residing in GPU memory can be distinguished from media data in system memory by "artificially" assigning different formats ("artificial" because both may technically represented using the same bit pattern) with conversion filters to transfer data back and forth. It then becomes the filter graph manager's responsibility to choose different filter nodes depending on the filter sink (or, alternatively of the pins to agree on either format during format negotiation).

In a sense, the filter graph manager may perform a limited form of "global optimization" when constructing the filter graph. the optimization is necessarily static as the filter graph can not always be reconfigured at run-time since the filter nodes may have built up considerable state information that cannot easily be transferred to (or recovered by) newly added nodes. However, this optimization

## 2.1. MEDIA PROCESSING FRAMEWORKS

capability is indeed very limited as it must treat all nodes already introduced into the graph (e.g. to realize video or audio effects) as "black boxes".

This can result in less than optimal processing behaviour if filters perform actions only "conditionally", i.e. sometimes passing the data through unmodified, while transforming it at other times. This limitation becomes very visible when considering the filter graph example shown in 2.5 – the filter graph executor cannot know when video C fully occludes video B and thus inhibit further processing of B. While the provided service to automatically generate filter graphs from the intuitive timeline representation is certainly nice, it begs the question if a more procedural interface to perform the editing operations might be preferable to begin with.

### 2.1.3 Network Integrated Multimedia Middleware

Network Integrated Multimedia Middleware [40] (abbreviated as *NMM*) originates from a research project initiated at University of Saarland in 2000. It is intended to provide an architecture for distributed processing of multimedia data on multiple networked computers. It is neutral to the platform both the controlling application as well as the participating media processing nodes run on, and also supports a heterogenous mix of systems. While it operates on a significantly higher level than the previous two frameworks, it is included in this discussion as an example of an approach to distributed multimedia processing.

#### 2.1.3.1 Distributed processing concepts

In contrast to the previous two multimedia frameworks which concentrate purely on media processing performed on a single computer, NMM addresses the problem of distributing the processing to multiple nodes in the network. Like Direct-Show, NMM takes a flow-graph based approach to multimedia processing – the graph itself is constructed and maintained by the controlling applications, but it consists entirely of proxy objects that control the "real" filter nodes which in turn may be spread across the network.

While an existing distributed object framework like CORBA could have been used to base this design on, the NMM creators elected not do so for a variety of reasons (citing mainly efficiency aspects such as resource consumption, cf. [41]). Instead the proxy objects are neutral to the middleware they use for communication with their counterpart (they may e.g. encapsulate CORBA stub objects, or use a homegrown protocol).

The filter graph itself is composed of **Node** objects that represent a specific multimedia data transformation (cf. 2.1.2). Each **Node** may feature several input and output **Jacks** that represent the node's communication endpoint with a peer node. Both **Nodes** and **Jacks** are controlled by the master application through corresponding proxy objects (cf. figure 2.6).

# CHAPTER 2. RELATED WORK

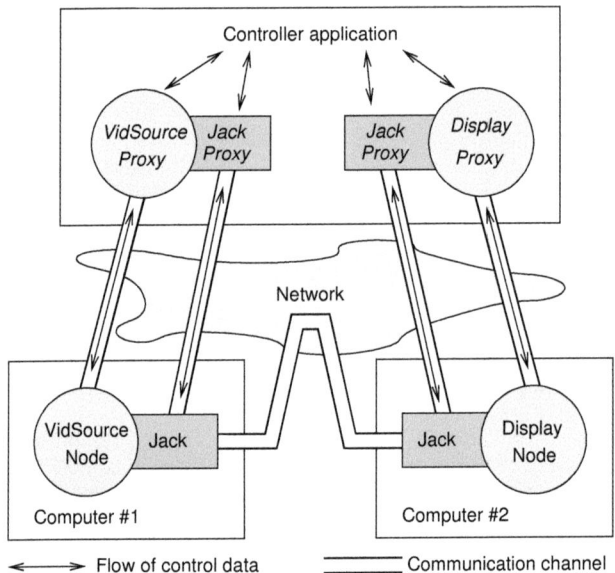

*The figure illustrates the distributed processing model of NMM: The application manages a logical filter graph consisting of proxy nodes, while the actual processing nodes may be spread across the network. The proxy nodes are the means by which the application can exert control over the processing nodes.*

Figure 2.6: Distributed filter graphs in NMM

The controller application is responsible for instantiating nodes (on the local system or on remote hosts), creating a filter graph and establishing the connection between **Jack**s that must exchange data. Instantiation of remote objects uses discovery and brokerage mechanisms very similar to those found in other distributed object systems, so it is not further discussed here. It should only be noted that NMM is also middleware-neutral in this respect and may thus in fact use several underlying mechanisms.

However, connection establishment between **Jack**s deserves further discussion, as the architecture treats communication channels as first class objects in the controller application: The controller chooses a means of communication to be used for exchange of media data between the **Jack**s, and then configures the **Jack**s accordingly (using the proxy object). In other words, the "negotiation" for a suitable communication channel is not performed by the **Jack**s or the physical machines they are hosted on, but by jack proxy objects in the controller application.

The ability of the controller to explicitly choose communication channels is essential also for other services provided by NMM (e.g. transparent migration of

## 2.1. MEDIA PROCESSING FRAMEWORKS

nodes while the filter graph is active).

### 2.1.3.2 Processing nodes

NMM itself defines a number of data formats as well as node types that can perform desired transformations and other media processing operations. Handling of (de)multiplexing, (de)compression and format conversions is similar to the strategy chosen in DirectShow.

NMM uses a unified component mechanism only for the proxy nodes contained in the controlling application's filter graph – the processing **Node**s that may be instantiated at remote on systems on the other hand do not have such a defined architecture. Instead, NMM takes the middleware-neutral approach also in this respect, and generally elects to implement processing **Node**s as wrappers around other existing media processing frameworks (like QuickTime or Direct-Show).

### 2.1.3.3 Discussion

NMM provides a highly abstract multimedia processing framework featuring a processing concept that is very similar to DirectShow. The architecture is considerably less refined than DirectShow, which is however more owed to constraints on developer resources than conceptual limitations.

Several of the comments given already in section 2.1.2.6 about filter graph based approaches in general do therefore also apply here. Mainly the "black box" nature of the filters can be a considerable cause of inefficiencies in that the graphs constructed are less than optimal for the desired processing.

The crucial difference to DirectShow that makes NMM quite remarkable is the separation of the *logical* and *technical* processing filter graphs: The logical filter graph consists of nodes that represent the controlling application's wishes, but it consists entirely of proxy objects that cannot actually provide the service they represent themselves. The technical filter graph consists of the nodes capable of performing the desired operations.

While the architecture strives to be neutral with respect to underlying middleware used for communication, this neutrality also has its downsides as it tends to limit the semantics to the lowest common denominator. Several services offered by sophisticated distributed object environments such as authentication are not considered at all. Unfortunately, experience tells that grafting such features later on a fully developed system tends to be rather difficult[10].

Interestingly, NMM assumes all participating nodes to have synchronized clocks (using e.g. NTP) despite its attempts to be independent from other middleware in general.

---

[10]e.g. authentication between remote nodes will require some form of credentials passing for impersonation

## Summary

The presented media frameworks exhibit different approaches to representing media data and media processing: From the relatively low abstraction level and high degree of control offered by QuickTime to the relatively shielded and rigid filter graph based models offered by DirectShow and NMM.

While the filter graph based approach can cope easily with very generic "temporally local" compressed media representations (cf. sections 1.4.1.4 and 1.4.3.3), QuickTime cannot really offer a sensible interface here as too much detail about the underlying media format is exposed and must be understood by the application – the option of falling back to the more convenient **MediaHandler** interface incurs the same semantic loss as the filter graph approaches.

It is also interesting to note that all presented systems are built on the assumptions that media is always represented in a time- and space-discretized format. While this limitation is not particularly severe today as this covers the vast majority of existing media data, this assumption is slightly at odds with trends towards "synthetic" representations (in the form of compositing or scene description operations) for which discretization is neither required nor a particularly natural representation.

The important observation is that the identified weaknesses are in fact architectural and not quality of implementation problems.

## 2.2 Media processing in the Linux environment

This section gives a brief overview of media processing facilities usually found in the Linux operating environment. For the most part these are small libraries that are very focussed on a single aspect instead of fully integrated frameworks, however two exceptions should be mentioned.

First, while NMM is not really targeted at fulfilling this purpose it could conceivably be used as a network transparent multimedia framework in this environment. However, NMM's approach of trying to be independent from the underlying target platform through its abstraction layers results in rather poor integration with the rest of the desktop environment – since it tries to assume basically nothing of the underlying platform in terms of services offered, it ends up duplicating a lot of infrastructure that is already in place. Additionally, lacking features such as authentication and an impersonation model for code executed on remote nodes makes it unsuitable for deployment in typical desktop scenarios where the network and other computers must be considered untrusted.

Second, *gstreamer* provides a cross-platform streaming media solution that is based on the same generic filter graph processing concept as DirectShow (it is therefore not separately discussed above). It is based on the **GObject** model developed as part of the Gnome desktop environment and retains close ties to its origins. Like DirectShow, *gstreamer* does not distinguish between logical

## 2.2. MEDIA PROCESSING IN THE LINUX ENVIRONMENT

and technical filter graph (cf. section 2.1.2.6) and as a result does not offer any support to delegate part of the media processing to remote nodes in the network (unlike NMM). This is unfortunate as most desktop environments support the network transparency offered by the underlying X Window System (see section 2.2.1.3) rather well. All other general remarks about filter graph based approaches apply here as well.

While *gstreamer* is widely deployed (and functionally compares quite favorably with QuickTime and DirectShow – note that these do not offer any network transparency either) this means that at the time of this writing none of the candidate multimedia frameworks can currently be considered suitable for the Linux environment without compromising the network transparency.

### 2.2.1 Low-level data capture and playback

#### 2.2.1.1 Video capture and overlays

Video for Linux (V4L) defines a kernel-provided interface that applications can use to gain access to framegrabber-like devices capable of performing direct memory access to the system memory. This includes basically all TV tuner cards and all non-USB and non-Firewire (see below) video acquisition devices, as well as a number of video overlay drivers. The original V4L interface is generally referred to as as V4L1 (Video for Linux version 1) while at the time of this writing most device drivers have been updated to the revised V4L2 interface.

The interface is very low-level and generally exposes only capabilities directly available in hardware – there is no software emulation of unsupported features requested by applications. Accordingly, applications can expect the operations to be very efficient but must on the other hand be prepared to be content with the restricted set of capabilities offered by some device. Many applications therefore do not use the kernel interface directly but make use of wrapper libraries that abstract and emulate access if desired, however no single standard library has emerged for this purpose (as the task of these libraries is in fact quite trivial). The V4L2 interface abstracts only the image capturing or overlaying capabilities of the underlying device – audio capture and playback is performed through the ALSA interface instead (cf. section 2.2.1.2).

For V4L2 devices the kernel exports a device node through which a descriptor to access the device can be instantiated. All further device access is channelled through this single descriptor which applications use to query capabilities, configure parameters and exchange data. The interface also provides synchronization capabilities through the usual unix file descriptor notification mechanisms (e.g. `poll` or asynchronous signals).

Devices can offer two types of interfaces for data exchange with the application (though no device is required to support both interfaces): One interface based on streaming `reads` (or `writes` for overlay interfaces), and one interface based on a shared memory ring buffer. The streaming interfaces are easier to use, each single `read`/`write` operation transfers one frame (or field) worth of data in the format negotiated between application and driver. While simple, this interface is prone to undetectable loss of frames (and therefore synchronization problems) as it does not support to exchange timestamps or other meta-data associated with the frames or fields.

The ring buffer interface requires the application and the driver to agree on a memory area that will hold multiple buffers for one frame or field each. The driver then provides an interface that allows to hand over buffers between driver and application – the producer will hand over the buffer after it is finished writing data, while the consumer will return the buffer after it has finished processing the data. Timestamps written by the data producer allow the partner to detect over- or underruns, it is up to the application to configure a sufficiently large number of buffers if it wishes to reduce the probability of this happening (at the cost of higher memory usage).

USB and IEEE1394 ("Firewire") devices are generally controlled using direct protocol access to the underlying bus exposed by the kernel, so the abstraction layer for these devices is not provided by the kernel itself but instead by user-space libraries such as `libavc1394`. For this type of devices the V4L2 API generally does not make much sense as these devices are technically incapable of DMAing complete frames from/to user-specified memory without further software assistance. Instead, interaction more resembles a network protocol exchange, thus the Linux designers elected to expose protocol-level access to these devices and therefore move the abstraction layer out of the kernel[11].

#### 2.2.1.2 Audio capture and playback

The Adavanced Linux Sound Architecture (ALSA) [66] provides applications with access to audio devices (most importantly PCM capture/playback and device control) on Linux systems. It operates on the lowest, technical level of audio playback and capture, allowing applications to transfer sample data to/from audio devices and receive transfer notifications (that can also be used for synchronization purposes). The architecture essentially consists of the following two parts:

- *Hardware drivers* realized as kernel modules; the drivers provide a common kernel/user-space interface that unifies access to different types of audio hardware.

- the *ALSA library* that acts as a shim layer between application and (hardware) drivers.

The ALSA library provides an interface that reflects the low-level hardware driver interface, but virtualizes the access such that the technical mechanisms

---
[11] It should be noted that several older drivers do in fact emulate the V4L interface, but this is gradually being deprecated in favor of more low-level access methods.

## 2.2. MEDIA PROCESSING IN THE LINUX ENVIRONMENT

of access to audio devices can be changed underneath. It provides several plug-in interfaces – e.g. it allows "virtual" PCM devices to be created through the "I/O plugin" mechanism (in fact, direct hardware access is realized through the built-in "hw" component).

PCM samples for playback and capture are stored in buffers of configurable size which are further subdivided into a configurable integral number of *periods* (buffer and period size are fixed before the playback or capture operation starts and may not be modified until it is finished). Applications are expected to fill/drain the buffer sufficiently frequently as to avoid under- and overruns. ALSA supports notification of the application as playback/capture of the next period starts, allowing the application to synchronize on the playback/capture rate.

Typically, ALSA maps the memory buffers from/to which audio devices transfer sample data using DMA into the application's address space; this enables applications to modify any sample value just after/until it is physically transferred from/to the audio device and thus allows good latency behaviour (essentially only limited by the real-time characteristics of the operating system and the audio hardware).

ALSA has emerged as the de-facto standard API for audio programming in the Linux environment. Detailed information including an API reference can be found in [1].

It should however be noted that other audio APIs exist that are less frequently used and normally layered on top of ALSA[12] as the abstraction layer offered by ALSA is quite low. None of these higher-level APIs has however received sufficiently wide spread adoption that they could yet be considered a standard API, the only real contender is `libsydney`.

### 2.2.1.3 Display

Linux provides several means to interact with the display subsystem, the method most prominent method is through the *X Window System* [55].

The X Window System distinguishes *X clients* – which are essentially the applications containing all processing logic – and the *X server* – which draws and displays graphics on behalf of the clients and manages input devices. Clients and server communicate through the X protocol which can be tunnelled through several transport protocols such as unix domain sockets (classical inter-process communication using bi-directional pipes) or TCP connections. The system is thus network transparent and allows to interact with applications on remote computers through a graphical user interface on the local terminal. Furthermore, it is extensible, allowing custom functionality to be added to the protocol in an upward compatible fashion.

---

[12] One notable exception is JACK when used in conjunction with several IEEE 1394-based audio devices as prefered by many music professionals: The devices are generally accessed using direct IEEE 1394 protocol access by JACK and audio synthesizing applications, bypassing ALSA.

At its core, the system provides mechanisms that allow clients to create, interact, and receive event notifications from various classes of server-side objects (in the terminology of X they are usually referred to as "resources"). The system provides a standardized way to identify these objects through 32-Bit numbers ("XIDs") in the protocol. All functional services provided by the X server use this infrastructure and provide objects with different capabilities, such as on-screen *windows*, off-screen *pixmaps*, graphics contexts or fonts. Clients initiate operations by sending a "request" containing desired parameters (including any server-side objects to be used in the operation) to the server, which will execute the command on behalf of the client.

X provides a hierarchical window model, with the whole screen represented as the "root" window and child windows being rectangular slices out of their respective ancestor window(s). Commonly, each application will create its own window as "virtualized screen area" to present its graphical interface. The system delegates most of the management tasks concerning the visual appearance and behaviour of the screen to the client applications – it is up to the clients to determine spatial positioning of the windows and determine what is drawn there. It is customary to centralize the positional management of windows in a single dedicated client application called the *window manager*.

Recent implementations of the X Window System provide two different drawing models: The legacy "core" X drawing model offers a rich set of geometric primitives (lines, polygons, ellipses) but is conceptually limited to drawing sharp-edged figures. The newer model based on the RENDER extension [45] is limited to relatively simple geometric shapes (triangles and trapezoids) and thus requires applications to perform tesselation of more complex shapes – but it offers a rich compositing algebra where geometric shapes typically serve as implicit masks with smooth boundaries, such that RENDER offers vastly expanded drawing capabilities much better suited for typical use-cases (where complex shapes like ellipse segments are indeed rather hard to find).

The newer rendering model is centered around the concept of a *Picture*, an abstract rastered image (see section 1.3.4.1) that may potentially include an alpha channel. The supported drawing operations combine a *source picture* clipped by a *mask* with a *target picture* using a compositing operator such as OVER. The mask may either be implicitly defined through a geometric shape, or explicitly given by another *Picture*.

While the X Window System supports a powerful drawing model, it is sometimes inevitable that the applications must transmit pre-rendered content to the X server. This is of course conceptually supported by the model, but the protocol does not support mechanisms to transmit compressed data – applications operating on the same physical machine have the option of exchanging data using shared memory segments, but for networked operations they are essentially restricted to "small" and/or infrequent image uploads if they want to maintain good performance.

It should be noted that on typical installations the X Window System is *not* the only avenue applications have of interacting with the display system – especially for games or very demanding graphics programs it is very common to require direct hardware access. The X server also provides some support for this model: It provides a mechanism to negotiate direct hardware access as well as means for coordination.

## 2.2.2 Media processing tools

### 2.2.2.1 Cairo

`cairo` is an abstract vector graphics API that supports multiple backends to perform drawing operations [11], including the X Window System (using the RENDER extension [45]), PostScript or Portable Document Format files, Apple Quartz, Windows GDI+ or OpenGL (through Glitz [43]).

The exported interface provides *surfaces* (represented as **cairo_surface_t**) to represent targets of drawing operations and *contexts* (represented as **cairo_t**) that represent the current drawing state (contexts are always assigned to a specific target surface). Additionally, it uses *patterns* as a generalized form of read-only images (represented as **pattern_t**) that can be applied to a target during drawing. Formally, each drawing command specifies an operation of the form

$$surface_{n+1} = surface_n \text{ op } (pattern \text{ IN } mask) \qquad (2.1)$$

where

- $surface_0$ is a fully transparent image

- $surface_n$ is the original image before execution of the $n$th compositing operation

- $surface_{n+1}$ is the image after execution of the $n$th compositing operation

- *pattern* is either a uniform color (possibly including transparency), a gradient, or an image itself (possibly modified by an affine transformation)

- *mask* is an opacity mask defined by a geometric shape; the shape may be a stroked or filled path; the path is either a sequence of straight lines and bezier splines, or the outline of a sequence of glyphs

- IN is the IN compositing operator and

- op is any of the compositing operators discussed in [50] such as IN, OVER or OUT, cf. section 1.2.2. (In fact, `cairo` supports a proper superset of the classical compositing operators).

```
void diagonal_blend(cairo_t *target, cairo_pattern_t *src1,
  cairo_pattern_t *src2, int width, int height, double t)
{
  cairo_pattern_t *gradient;

  /* create blend mask */
  gradient=cairo_pattern_create_linear(0, 0, width, height);
  if (t>0) cairo_pattern_add_color_stop_rgba(gradient,
    0 /* stop */, 1 /* R */, 1 /* G */, 1 /* B */, 1 /* a */);
  cairo_pattern_add_color_stop_rgba(gradient,
    t /* stop */, t /* R */, t /* G */, t /* B */, t /* a */);
  if (t<1) cairo_pattern_add_color_stop_rgba(gradient,
    1 /* stop */, 0 /* R */, 0 /* G */, 0 /* B */, 0 /* a */);
  cairo_set_source(target, gradient);
  cairo_pattern_destroy(gradient);
  cairo_set_operator(target, CAIRO_OPERATOR_SOURCE);
  cairo_paint(target);

  /* filter second image with mask */
  cairo_set_source(target, src1);
  cairo_set_operator(target, CAIRO_OPERATOR_IN);
  cairo_paint(target);

  /* blend masked second image over first image */
  cairo_set_source(target, src2);
  cairo_set_operator(target, CAIRO_OPERATOR_DEST_OVER);
  cairo_paint(target);
}
```

*The function takes a given drawing context* target *(which must previously have been instantiated for a target* cairo_surface_t *and performs operations on this drawing context that blend the two given images. The parameter* t *is used to indicate the "progress":* t=0.0 *results in the image* src1 *being visible,* t=1.0 *will yield the image* src2. *Intermediate values result in a smooth diagonal blend of the two images. See figure 2.8 for an illustration of the effect.*

Figure 2.7: Diagonal blend transition between two images

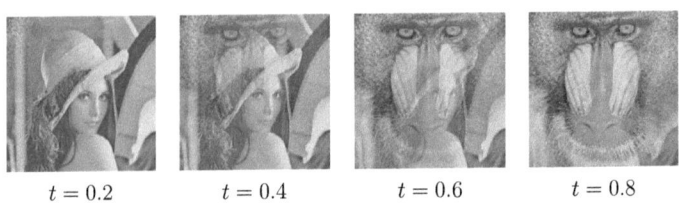

$t = 0.2$     $t = 0.4$     $t = 0.6$     $t = 0.8$

*Diagonal blending of two images, using different values for the progression parameter t. See 2.7 for the code generating this effect.*

Figure 2.8: Diagonal blend transition between two images

## 2.2. MEDIA PROCESSING IN THE LINUX ENVIRONMENT

Drawing contexts store all information contained on the right hand side of equation 2.1, i.e. the pattern, the mask as well as the compositing operator. cairo_t context objects are stateful to allow incremental construction of paths that delineate a mask, and it supports affine-linear coordinate transformations of patterns – the drawing model is in fact similar to PostScript in that it provides methods to save and restore drawing states.

Figure 2.7 shows an example of using cairo to composite two given images: The first block of calls creates an alpha mask that transitions from fully opaque in the upper left to fully transparent in the lower left. The parameter t controls the position and alpha value of an intermediate point on the connecting line; varying the parater from 0 to 1 in time allows to realize a simple blend effect. The further calls combine the mask with the second image and blend the result over the first image. See figure 2.8 for an illustration

The drawing model supported by cairo maps well to the RENDER extension of the X Window System – equation 2.1 essentially also describes the X rendering model. However, X supports only rather primitive shapes and has no concept of a drawing context, so cairo must support several additional services such as tesselation, state management or caching of resources used for scratch intermediate pictures that are used as patterns.

Cairo is poised to become the de-facto standard API for 2D graphics in the Linux environment, however it also enjoys considerable popularity on other platforms it supports, in particular for embedded systems. While cairo can conceptually support modular backends, the backend interface is at the time of this writing not finalized yet and thus officially unsupported. For more information refer to [10].

### 2.2.2.2 Codec libraries

The Linux operating environment usually includes several different libraries providing compressor and decompressor implementations for various compressed media representation formats (in the sense of section 1.4). On the one hand there are libraries concerned with individual formats, such as libjpeg and libpng for images or libvorbis for audio. Usually, their purpose is to provide comprehensive support capabilities of the underlying format. They do not provide a common API but instead prefer to expose all peculiarities of the format. Due to their completeness they can usually be considered the de-facto standard libraries for the specific media representation.

The second group of libraries are "codec collections" – their goal is to provide support for many different formats through a common API, and therefore necessarily hide many of the individual format's characteristics. Many of these libraries have their origins in different "media player" projects, one particularly popular example is the ubiquitous vlc media player (but interestingly none of these projects has considered the issue of network-transparent playback). This heritage also means that most of these libraries follow a "streaming" processing paradigm.

The libraries vary wildly with respect to format coverage. One particularly comprehensive collection is provided by the ffmpeg project: It provides a command-line tool for encoding and transcoding media [65]. It makes use of two libraries, libavcodec and libavformat that form part of the distribution – these provide audio and video codecs as well as support for several types of container formats, respectively.

libavcodec provides a unified interface for video and audio codecs as well as a plugin infrastructure for registration, lookup and instantiation of encoder and decoder instances. The instances are functionally roughly equivalent to DirectShow compression and decompression filters: They process data according to the streaming paradigm, they are stateful and transparently perform internal buffering and reordering to cope with out-of-order en- and decoding.

# Chapter 3

# Media processing framework architecture

This chapter outlines the design of the media processing framework, and in particular the core media toolkit library – imaginatively dubbed libmedia – providing the interfaces and basic functionality. In the first part the "guiding ideas" underlying the architecture will be introduced and motivated in front of the background of the findings in section 2.1. These ideas are realized in the architecture described in the succeeding sections.

Section 3.2 introduces the core, media-type independent concepts. This includes the base classes that provide the time concept, modularization and the programming interface concepts used for I/O of media data. Section 3.3 discusses how support for the most commonly used types of media (audio, images, video) is realized on top of the core abstractions; in the architecture to be presented media processing is inseparably linked to media representation, so this aspect is also covered here. Section 3.4 discusses capture and rendering of audio and video media; the concept of "rendering" is of particular importance in this architecture – it is both much more generic than "playback" and considerably more complex than in other media architectures. Section 3.5 finally discusses the "document" concept used to represent persistently stored media; typically this is done in container files, but the document concept is more generic.

## 3.1 Design choices

The purpose of the multimedia framework as a whole is to assist application programmers in realizing their application's media processing requirements. The key observation is that application writers generally wish to formulate media processing intent in terms of steps to be performed on the *media content* (e.g. "darken this image") without caring about the concrete *technical representation* (e.g. color model) currently used for the media element. This naturally leads to a number of abstractions that should be provided by any media processing framework, such as *images, video* and *audio sequences* as well as multimedia *container documents* (cf. QuickTime movie files, section 2.1.1.2.).

While these abstractions themselves are near-universal, the model provided by the media processing framework for manipulating these data elements is not. The following sections will motivate and present the design choices made for `libmedia`.

### 3.1.1 Processing model

While filter graph based architectures like DirectShow provide an easily understandable mode of operation, they have several weaknesses (as pointed out in section 2.1.2.6 already):

- The filter graph approach forces the application to adopt an event-driven programming paradigm as the processing is controlled entirely by the filter graph executor. While this may be an acceptable restriction for GUI-driven interactive applications (that are generally structured for event-driven processing anyways) this is highly undesirable for applications that are essentially data-driven (such as batch processing) or can for other reasons not tolerate the inversion of the control flow (which is a typical problem for many network server applications).

- Since the modularization concept requires the filter nodes to be treated as "black boxes" this can prevent interesting optimizations, e.g. if two chained filters perform inverse operations: While it would be desirable to simplify the graph by eliminating the two nodes, this would require knowledge of the inner workings of the two nodes – which however is not available without compromising on the modularization concept.

- The stateful nature of the filters makes it difficult or even impossible to reconfigure the graph at run-time – unconditionally safe modifications can only affect nodes that are known to be stateless.

- A second ill effect of the state implicitly held by filter nodes is that the considerable loss of semantics as the graph represents only the "forward" processing of the data: Data received by the "sink" nodes of the graph cannot be correlated back with data pushed by the "source" nodes into the graph.

- For complex media processing operations, "time lines" editing provide a considerably more intuitive representation. While transformation of the time line editing into a filter graph representation is possible, the more obvious realization would be to iterate through the time axis and perform required transformations procedurally.

These weaknesses make an imperative programming paradigm as featured by QuickTime appear preferable. Moreover, if media processing operations are

3.1. DESIGN CHOICES                                                                     83

realized as procedure calls, it is trivial to "wrap" these calls into filter node instances and thus create a filter graph processing concept on top of the underlying procedural interface – however the opposite is not true, so an **imperative programming paradigm** is the more generic of the two approaches and has been chosen for this architecture in favor of filter graph processing.

### 3.1.2 Data model

In the summary on page 72 of section 2.1 it was pointed out that the media processing frameworks discussed there universally assumed temporally and spatially discretized media representations. Considering the introduction given in section 1.1 this limitation appears arbitrary and unnecessary: While discretized representations play an important role in practical applications, they are better regarded as imperfect approximizations of conceptually continuous time and space domains. This becomes particularly obvious if the data must be "resampled" either spatially or temporally: Lacking a canonical and theoretically sound model how the discrete data is to be interpreted if interpolated, resampling is not well defined.

Therefore, for this architecture media will be treated as **time- and space continuous** (where applicable for the particular media type in question). Moreover, discretized media will be treated as special cases of continuous media with canonical interpolations such as described in sections 1.3.2.1, 1.3.4.1 and 1.3.5.1 for the media types discussed there. In keeping with object oriented design principles this means that the architecture must provide a base abstraction for the most generic (continuous) representation alternative, while the more specialized discretized representations inherit from the generic one.

With the architecture providing these abstractions this also allows to address an issue brought up in section 2.1.1.7: While QuickTime **Movie** objects allow to retrieve media *data* for a given image, it requires the caller to resolve (recursive) decoding dependencies itself. The provided abstractions should therefore include any required *meta-data* (like implicit dependencies) for unambiguous interpretation alongside the actual data.

### 3.1.3 Execution model

Section 2.1.1.7 pointed out that the "immediate" execution model featured by QuickTime components leads to difficulties in delegating processing which is however essential for this architecture as network transparency is one of the main development incentives. A similar situation also exists in DirectShow (cf. discussion in section 2.1.3.3) as the lacking separation of logical and technical filter graph presents an obstacle for *dynamic* delegation (though it can still realize *static* delegation at filter graph setup time using the format distinction explained in 2.1.2.6).

NMM addresses the issue of delegation by introducing an additional abstraction layer through proxy objects for the filter nodes, i.e. for the *processing* objects. The same solution can however not be applied here: In case the operation is delegated to a remote processing object, the imperative immediate execution paradigm would just incur useless communication overhead as the data must be transferred to the processing object and back eventually.

What the architecture therefore elects to do is to introduce an additional abstraction layer on the *data* objects instead, and to give up the immediate execution paradigm in favor of **retained-mode processing**: Instead of applying transformations on media elements, a description of how the media elements should appear is built up. A bridge towards the imperative programming model is provided by additionally introducing the concept of **lazy evaluation**: Any transformation that the application wishes to apply to any media data element is added to the description of the media element (and thus queued up for execution at a later point in time) – this allows to delegate the complete sequence of all required processing steps to be executed as a single entity, eliminating the need to communicate intermediate results between the steps. As will be discussed later in more detail (section 3.4.3), the retained-mode processing paradigm enables a number of interesting optimizations by the media architecture.

### 3.1.4 Format transformations

The filter graph based approaches from section 2.1 feature a mechanism to automatically complete a given, as of yet disconnected graph by introducing auxiliary format conversion nodes to make the graph connectable. This service is convenient for application programmers as conversion of media data of an abstract type (e.g. an image) into an arbitrary representation of the same abstract type (e.g. a rastered image in a specific color model) is automatic.

The same convenience can be provided in an imperative programming model by making media elements **weakly-typed with implicit conversion** when a transformation is applied that assumes a specific format. In conjunction with the retained-mode processing paradigm explained in the previous section this of course makes the auxiliary format conversions delegatable as well.

### 3.1.5 Component and object model

All of the architectures discussed previously draw their extensibility from an underlying component mechanism – whether the component model is part of the media framework itself (e.g. QuickTime, see section 2.1.1.1) or part of the operating environment (e.g. DirectShow and COM+, see section 2.1.2.1) that is just reused.

The component models underlying the systems discussed are provided as *library functions* and not as *language features*. For example, the QuickTime component mechanism is not type-safe: As each component instance regardless of

## 3.1. DESIGN CHOICES

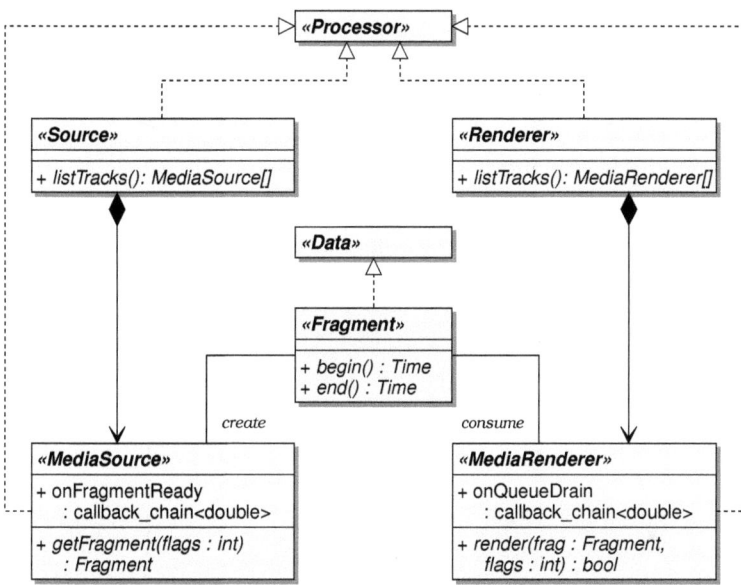

Figure 3.1: Classes providing basic media processing capabilities

type is represented as a **ComponentInstance** object, it is the caller's responsibility to only use methods that are indeed supported by the underlying implementation – the compiler has no mechanism to statically verify this as instances are indistinguishable at the language level[1]. The COM+ component model on the other hand requires "binding" through a particular interface using the **QueryInterface** method prior to calling using it – afterwards all method calls are made through a type-safe class interface.

This "library" approach to the component model leads to considerable functional duplication of services already provided by most modern run-time environments – Java and C++ for example provide language mechanisms to bind to a specific interface implemented by an object at run-time through the type-cast syntax (e.g. using **dynamic_cast**). In addition to being better known to programmers, these also offer the benefit of allowing better static checking through both the compiler or other verification tools.

For this architecture it has therefore been decided to **rely on the run-time environment for component services** (see section 3.2.1) – for all required ser-

---
[1]QuickTime supplies the **ComponentCanDo** function which may check at *run-time* if a given operation is supported by a component instance in question.

vices existing features of the language (classes, inheritance and run-time type information) or other parts of the run-time environment such as the link editor (symbol resolution, dynamic loading) are to be reused.

## 3.2 Core architecture

`libmedia` distinguishes two abstract classes that are the base of most other classes:

- **Data**: Designates objects that contain or represent any type of media data. This includes multimedia documents, individual media or parts thereof, byte-data that corresponds to a media element in a specific representation, or encapsulated operating system objects that can contain or otherwise represent media data.

- **Processor**s: Designates objects that consume, produce, transform, or otherwise manipulate **Data** objects, including accessors that store or retrieve data to or from containers.

Both are mutually exclusive, i.e. classes may never derive from both. **Data** objects generally encapsulate static system resources (see the implementation notes in appendix A.1 for a more detailed description of the data model), while **Processor** objects generally have no static resource usage (other than the memory used to represent the object structure itself). The framework provides an automatic garbage collection mechanism for all objects that derive from these base classes.

The library provides a number of pre-defined abstractions for commonly used media elements. The most important one is the abstract **Fragment** concept used for linearly time-dependent media: Each **Fragment** represents a short temporal interval of media. Two timestamps denote the beginning and end of the interval, instances of subclasses of **Fragment** will additionally contain the media snippet corresponding to the interval. **Fragment** cannot represent any media by itself and cannot be instantiated, derived subclasses represent fragments of specific elementary media types (see section 3.3).

Objects providing the **Fragment** interface are usually obtained through the **MediaSource** interface (**getFragment**) and are usually consumed by **MediaRenderer** (**render**), applications can however also synthesize or process fragments themselves. Like **Fragment**, **MediaSource** and **MediaRenderer** are abstract and cannot be instantiated, derived subclasses deal with specific media types.

Multiple **MediaSource**s can be aggregated through the **Source** interface, equivalently multiple **MediaRenderer**s can be aggregated through the **Renderer** interface (see figure 3.1 for the relationship of classes). **Source**s generally represent objects that are capable of producing *multi*media data, such as capture devices, receivers of network-transmitted media streams or read accessors into

## 3.2. CORE ARCHITECTURE

files, while **MediaSource**s represent the individual media delivered by the source (e.g. individual "channels"). The aggregating **Source** acts as control instance for all parameters that affect all subordinate **MediaSource**s collectively. This includes for example start/stop control or *synchronization*.

Equivalently, **Renderer**s generally represent objects that are capable of processing *multi*media data and transferring it to an entity external to the library – such as presentation devices, transmission via network connections or files. **MediaRenderer**s interpret given media fragments and transform the data into a form suitable for the target they represent. This does not necessarily mean that a **MediaRenderer** produces a *perceptible* representation of the media (e.g. a visible image) since **MediaRenderer**s are also used to represent write accessors into files.

No interpretation for the individual **MediaSource**s or **MediaRenderer**s grouped by a **Source** or **Renderer** instance is mandated through the framework – the application is free to treat them e.g. as "alternative" or "complementary" media channels, the creator's intention must be inferred from additional context.

The two "ends" of a media processing chain are not symmetric in this architecture (unlike filter-graph based architectures, cf. 2.1.2): Due to the retained-mode processing paradigm (see section 3.1.3) it is in the **MediaRenderer**s that all processing is performed. To reflect this asymmetry the consumers have been named *renderers* instead of sinks to emphasize this difference.

Both **Source**s and **Renderer**s can indicate that they operate in real-time. For **Source**s this means that fragments become ready for reading at some rate (and the application must read them periodically to avoid overruns), for **Renderer**s this means that they expect to periodically receive fragments (and applications must provide them to avoid underruns).

### 3.2.1 Modularization and component model

Like the reference architectures presented in chapter 2, `libmedia` features a modular approach to providing functional services. This is generally realized as name-to-object in conjunction with the *abstract factory pattern* (e.g. [17] pp. 87ff): For example, the architecture defines the **MIMEHandler** interface (section 3.5.2) which provides as its main service the **createDocumentFromFile** method to instantiate objects that can process a file of a specific MIME type. The registry is then used to associate the name `video/quicktime` with a single object that acts as factory to create accessor objects to QuickTime movie files through an overridden **createDocumentFromFile** method.

The approach taken for `libmedia` however differs slightly from other component models: QuickTime e.g. provides an abstract **Component** interface as base class for the **ComponentInstance** factories; This means that applications utilizing the component manager (fulfilling the role of the single registry) are responsible for ensuring type safety for the instances themselves.

Instead, `libmedia` provides multiple, but type-safe registry lookup functions – in the example above, the lookup function will always return objects of type **MIMEHandler** (instead of some "component" supertype) which the application can use with compile-time type checking.

The name registry is realized by mapping names to the identifier namespace of the run-time environment[2] and using the introspection capabilities for symbol resolution (i.e. in the above example the name resolution process would look for a global object instance by the name **media::mimehandlers::video::quicktime**). This has the advantage that an application can either use run-time symbol resolution through the lookup-function if the name of the required component is not known at compile time, or it can directly reference the factory object if a specific component is absolutely required in a specific place (which will consequently also result in a compile- or link-time error in case the component is unavailable).

The individual registries are organized as sub-namespaces of the global identifier namespace. In particular, they are not declared in advance and individual subsystems may introduce and use their own registries for internal purposes (e.g. the QuickTime and AVI file handlers use the **media::quicktime::trackhandlers** and **media::avi::trackhandlers** namespaces as registries to assign compressed media formats to the four-letter codes used in the file formats[3]). This also means that registration of new components is done by simply declaring a global object with appropriate name.

Provisions are in place to also handle dynamically loadable modules transparently – symbols not found by introspection of the executable will simply be looked for in the candidate loadable modules, with an optional symbol cache to speed up lookup and browsing. A technical description of how this approach can be realized in the target Linux/ELF environment is given in appendix A.4.

### 3.2.2  I/O model

Both **MediaSource**s and **MediaRenderer**s can operate in either *blocking* or *non-blocking* mode. Applications may choose *non-blocking* mode of operation when **MediaSource**s/**MediaRenderer**s are real-time and must choose *blocking* mode otherwise.

In *blocking* mode a call to retrieve a **Fragment** from a **MediaSource** will block until at least one **Fragment** is available; thus progress of the application will implicitly be flow-controlled by the **MediaSource**. Equivalently, handing a **Fragment** to a **MediaRenderer** may block while the **Renderer** is busy processing a backlog of **Fragment**s[4] and may thus exercise implicit flow control as well. *Blocking* mode of operation is suitable if the application either wants to perform "bulk" processing

---
[2] In some cases this requires name-mangling as not all characters are allowed in identifiers.

[3] These distinct registries account for the fact that the four letter codes have different meanings in the different container formats – e.g. motion jpeg is identified as `'mjpa'` in QuickTime files, but as `'MJPG'` in AVI files.

[4] Note that *non-blocking/blocking* and *asynchronous/synchronous* processing are two orthogonal concepts!

## 3.2. CORE ARCHITECTURE

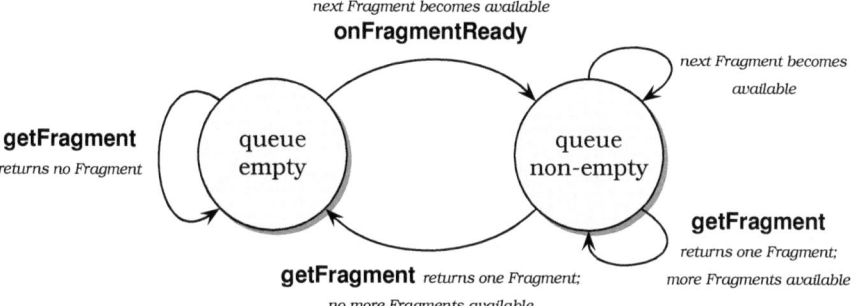

The diagram depicts the state transitions initiated by non-blocking calls to **getFragment** or availability of new media Fragments, and illustrates when **onFragmentReady** callbacks are delivered

Figure 3.2: State transition diagram for non-blocking access to **MediaSource**s

of media data (and thus does not care about timing at all), or can for external reasons be confident that blocking does not interfere with timing requirements (e.g. the application may assume that reading media data from/writing media data to disk can be performed significantly faster than real-time, or it may employ multiple threads that may individually block waiting for data).

In *non-blocking* mode reading a **Fragment** from a **MediaSource** or writing a **Fragment** to a **MediaRenderer** will never block, but may instead inform the application that the operation cannot currently be executed (e.g. because no **Fragment** is currently available from a **MediaSource** or a **MediaRenderer**'s backlog queue has reached its limit). Flow-control is in this case exercised explicitly through two callback chains (**MediaSource::onFragmentReady** and **MediaRenderer::onQueueDrain**).

Registered callbacks are delivered to notify the application of state transitions between "fragment queue is empty" and "fragment queue contains at least one element" for **MediaSource**s (cf. figure 3.2), and "render queue is full" and "render queue has space for at least one fragment" for **MediaRenderer**s (cf. figure 3.3). Notification is "edge-triggered" to inform about transition between the two respective state for performance reasons, as this mechanism is only intended to provide an "on/off" flow-control mechanism to detect and avoid buffer over- or underruns in exceptional situations (e.g. loss of synchronicity due to run-time errors). Applications that want to perform real-time media processing are generally advised to employ the mechanisms outlined in the following sections for proper synchronization.

Blocking and non-blocking modes of operation correspond exactly to the Posix blocking and non-blocking I/O concepts. Non-blocking I/O is provided to acco-

90      CHAPTER 3. MEDIA PROCESSING FRAMEWORK ARCHITECTURE

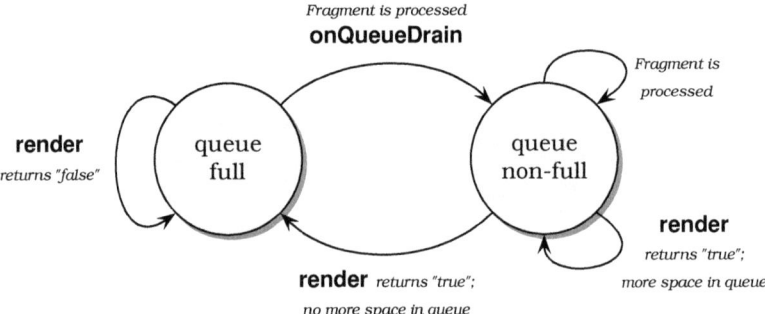

The diagram depicts the state transitions initiated by non-blocking calls to **render** or processing of one fragment, and illustrates when **onQueueDrain** callbacks are delivered

Figure 3.3: State transition diagram for non-blocking access to **MediaRenderer**s

modate reactive (event-driven) applications.

### 3.2.3  Time model

As indicated in section 3.1.2, time is treated as a continuous scalar variable (in contrast to other media systems where time is always discretized into "ticks"). This approach relieves the programmer of considerable work as events may be associated to arbitrary points in time. However, this requires additional architectural support for sources of time information that provide only infrequent timer "ticks".

The architecture allows applications to process media by receiving media fragments from sources, specifying transformations on the data, and finally handing the fragments over to renderers. This general order of operations is always the same and independent from the actual implementations of the participating components.

Both the sources of media fragments and the renderers may be subject to real-time requirements: For media sources this typically means that they correspond to data acquisition devices (such as video cameras, frame grabbers, audio digitizers) or real-time media transmissions (such as live broadcasts). For media renderers this typically means that they correspond to output devices (such as displays or speakers).

Real-time sources and real-time renderers must autonomously execute actions at defined points in time to perform their functions. A media processing chain as outlined above requires coordination between the different actors, and this section will present the mechanisms that enable this coordination.

## 3.2. CORE ARCHITECTURE

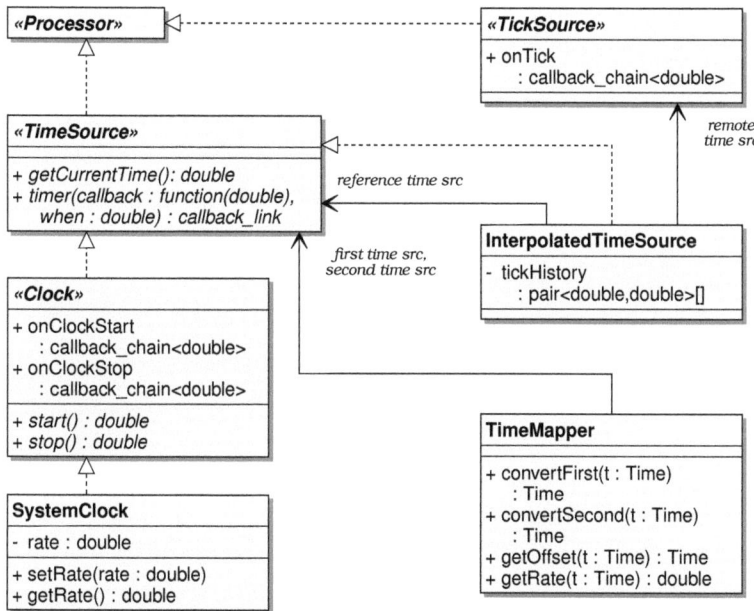

Figure 3.4: Relationship of time-related classes

### 3.2.3.1 Time representation and time sources

The timestamps contained in media fragments are interpreted by (at least) two different entities:

- Media data sources generate fragments and have to assign time stamps. Data acquisition devices will usually use the (idealized) point in time the data contained in the fragment was recorded.

- Media data renderers use the timestamps to decide on the temporal placement of media data; for output devices this usually determines the point in time the media element contained in the fragment is to be played back.

The different objects interpreting timestamps as points in physical time need a mapping between their logical notion of time and physical time (which will be referred to as *wall-clock time* subsequently). This mapping is provided through the **TimeSource** interface which offers only two types of services:

# CHAPTER 3. MEDIA PROCESSING FRAMEWORK ARCHITECTURE

- Applications can inquire the logical time represented by a **TimeSource** at the current point in physical time

- Applications can schedule callbacks to be executed at a given logical point in time[5]

These services allow applications to synchronize actions to individual internal time sources. Applications may only assume that the time represented by a **TimeSource** is monotonic with respect to physical time. Generally, the progression of time represented by a **TimeSource** is not under the control of the application (they may e.g. represent the oscillator of an audio device that cannot physically be influenced at all). Derived classes represent time sources that provide additional guarantees or provide additional functionality that allows more control by the application:

- **Clock** provides start/stop control in addition to the services provided by all timer sources. Clock objects usually represent physical clocks with fixed rate such as the clock signal used for audio DACs/ADCs ("fixed" rate in this context means that the rate cannot be changed without stopping/starting the clock, thus disrupting the flow of time).

- **SystemClock** represents a clock that relates to the system timer. In addition to the start/stop control provided by **Clock**, its rate may be varied[6].

Instances of **SystemClock** have a number of important properties. First, applications may assume that they can inquire the current time with very little and – most importantly! – *constant* overhead. (This is not generally true for all time sources as fetching the "current" time may e.g. involve network communication. Stated differently, applications may assume **SystemClock** instances to be jitter-free with respect to physical time).

Second, applications may assume that **SystemClock**s provide time information with "virtually infinite" resolution[7]. In practical terms "virtually infinite" resolution means the following: Assume that the same clock is sequentially inquired for the current time twice; assume further that both inquiries are separated by an action that requires progression of physical time (e.g. waiting for an event from a device); then the **SystemClock** will yield two different values[8].

---

[5]It should be noted that any real-time guarantees about timely execution of scheduled callbacks depend on factors external to the architecture description (such as the execution environment). Concrete implementations will usually only provide statistical guarantees as this is generally sufficient for multimedia applications.

[6]This does not imply that the system timer itself is reprogrammed at a different rate (although such an implementation would be permissible). Instead, the time value provided by the system clock may simply be transformed to match that of a virtual clock progressing at the desired rate.

[7]In practice, it is sufficient that the resolution be of the same order as the clock frequency of the processor executing the code (any higher resolution is useless for practical purposes). Thus the cycle counters of modern CPUs provide a suitable mechanism to implement **SystemClock**s.

[8]The prerequisite – that an action which takes a certain duration of physical time takes place in between – is very important. This allows an implementation to correctly yield the same value

## 3.2. CORE ARCHITECTURE

Third, all instances of **SystemClock** share the same time base. This means that the temporal relationship between two **SystemClock** instances is always precisely defined by their offset and relative rate of progression. Figure 3.4 summarizes the relationship of all time-related classes.

### 3.2.3.2 Ticks and tick interpolation

Some sources of time can not be explicitly queried for the current time (or it would be impractical to do so), instead they only provide infrequent "ticks" that mark individual points in time. Examples include communication over packet networks when each packet contains (or is implicitly associated with) a timestamp: Arrival of an individual packet marks a point in time (through the timestamp) but there is no well-defined "time value" in between the arrival of two packets[9].

These types of entities can therefore not directly usefully be represented as **TimeSource**s. They are instead represented as **TickSource**s and provide only one service: They can generate a callback for each individual tick. Each tick is represented as a single scalar value (like points in time are) that is passed to the callbacks. Ticks need not be equidistant, and neither need the tick callbacks be temporally equidistant.

In practice, **TickSource**s often represent clocks to which the application does not have direct access to, but that infrequently send "messages" informing about the current value of the clock. Examples include clocks in other networked computers (through packets received from a server that is sending data at a rate determined by its own clock), or physical clocks attached to the local computer that do not provide sufficient resolution to be adequately represented by the **Clock** interface explained in the previous section. This type of clocks will be referred to as *remote clock*s for the following discussion.

Let $t_k$ be the points in real-time (measured by an arbitrarily chosen *reference clock*) that a tick event is delivered and $\tau_k$ be the value passed as argument to the callback functions for each tick event. If the tick values correspond to "readings" of a *remote clock* roughly progressing at real-time, and tick callbacks correspond to "messages" communicating the tick values, then the following condition holds:

$$\tau_k \in [\alpha t_k + \delta_{min}; \alpha t_k + \delta_{max}]$$

where $\alpha$ denotes the rate of progression relative to the *reference clock*, and delivery of tick callbacks may be delayed by an unpredictable amount of time within the interval $[\delta_{min}; \delta_{max}]$. An application may wish to synchronize operations to the

---

twice if no such action has taken place on the grounds that an "infinitely fast" computer could have executed the instructions taking only an "infinitesimal" amount of time.

[9]If the underlying packet network supports isochronous transfers with hard real-time guarantees a useful time value in between two packets could in fact be derived. However, it would in this case be more practical to implement a **TimeSource** derived from the clock signal of the network interface adapter.

*remote clock*, and for this purpose the architecture provides a mechanism to construct a *local clock* that is *synchronized* to the *remote clock*; the local clock can then in turn be used to time operations accordingly. Ideally, the local synchronized clock would be a clock progressing at rate $\alpha$ with an appropriate offset with regards to the chosen *reference clock*, but given the data points $(t_k, \tau_k)$, the parameters $\alpha$ and $\delta_{max} - \delta_{min}$ can only be determined statistically[10].

The class **InterpolatedTimeSource** provides such a clock that is synchronized to a *remote clock* given only infrequent ticks provided through a **TickSource**. In particularly it

- determines the required parameters through statistical analysis of the available data $(t_k, \tau_k)$

- interpolates the time represented by the *remote clock* between two points $\tau_k$ and $\tau_{k+1}$ (hence the name)

The problem addressed by **InterpolatedTimeSource** is known as "external clock synchronization"; it has been covered extensively in the literature, and algorithms developed for this problem can directly be applied to implement **InterpolatedTimeSource**s. In particular, Schmid and Schossmaier address the problem of clocks with limited timer granularity in [56].

### 3.2.3.3 Relationship of multiple time sources

Applications may frequently communicate with multiple entities that use their own time source to provide timing (e.g. real-time data acquisition and presentation devices); if the time sources represent distinct *physical* clocks the application cannot assume that progression of time is the same for each[11].

Applications nevertheless need to coordinate such entities, and for this purpose the class **TimeMapper** provides a mechanism to determine the relationship between different time sources. Conceptually, the class allows an application to translate timestamps between two different time frames into each other. If both time sources relate to the same physical clock the mapping can be determined strictly arithmetically (this is e.g. the case for any pair of **InterpolatedTimeSource**s that use the same reference clock). Otherwise it must be determined by observing both time sources (see figure 3.5); in this case the mapping is generally subject to statistic errors.

Both strategies are encapsulated into **TimeMapper**, so that applications can be unconcerned if two different **TimeSource**s are synchronized, but the media processing framework will make best use of all information available to it. This

---

[10]In practice, the drift between two clocks may even be variable – this means that the requisite paramters $\alpha$ and $\delta_{max} - \delta_{min}$ need not even be constant, and the local synchronized clock must adapt to changing parameters.

[11]Due to manufacturing inaccuracies oscillators with the same nominal frequency will nevertheless diverge slightly; thermal noise adds additional unpredictable synchronization error.

## 3.2. CORE ARCHITECTURE

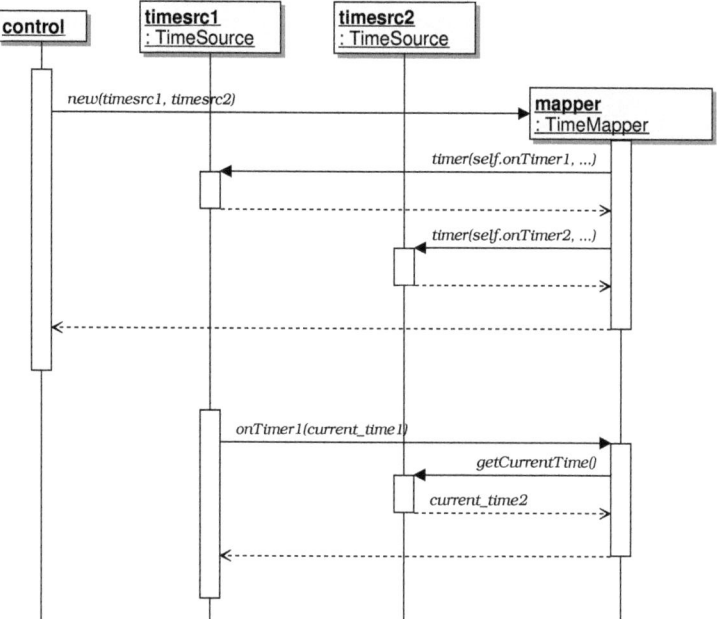

The **TimeMapper** determines the relationship between two time sources by periodically reading the current time of both sources. This is generally achieved using the callback mechanism offered by **TimeSource**

Figure 3.5: Mapping of timestamps between two different sources

approach is consistent with the general architectural decision to let applications express their "intent", while the media framework figures out the best mechanism to satisfy the application's requirements.

**TimeMapper** provides the following guarantees for the mapping of timestamps:

- *Invertibility*: **convertFirst** and **convertSecond** are mutually inverse.
- *Continuity*: The mapping is a continuous function.
- *Repeatability*: If conversion of a timestamp $t_{future}$ has been requested, then the mapping for all timestamps $\in [t_{now}; t_{future}]$ is fixed.

The last guarantee is important as the **TimeMapper** continually collects data from the two time sources and may detect drift that has to be corrected. This correction may change the mapping of future timestamps, but *repeatability* guarantees that the mapping for timestamps that have been converted already is not invalidated.

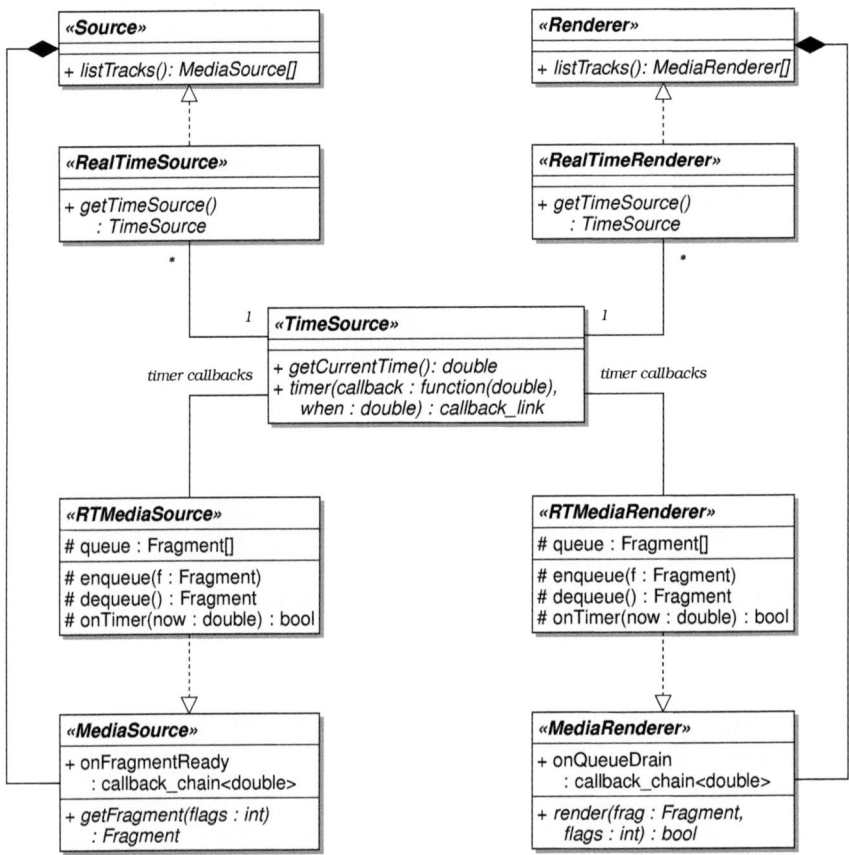

Figure 3.6: Real-time sources and renderers

## 3.2. CORE ARCHITECTURE

### 3.2.3.4 Real-time sources and renderers

**Source**s that deliver media data at a fixed rate that asymptotically approximates the timestamps of the media fragments are considered real-time sources. The rate at which accessors to documents stored as files read media data is determined by the speed at which applications request media fragments, thus flow control is implicitly exercised by simply fetching fragments slower or faster.

**RealTimeSource**s on the other hand expect the application to read data at *their* rate – flow control cannot be exercised implicitly by requesting data slower or faster. In fact, many real-time, sources (e.g. audio digitizers) cannot be flow-controlled at all. For those real-time sources that *do* support flow-control applications must explicitly request to speed up or slow down the data rate through secondary mechanisms.

Real-time sources provide a **TimeSource** to represent their flow of time. This allows applications to synchronize operations on sources through the mechanisms described in section 3.2.3.1. Additionally, **TimeMapper** allows applications to synchronize multiple sources or renderers by adapting timestamps of fragments accordingly.

In analogy to real-time sources the architecture provides **RealTimerRenderer**s that consume media data fragments in real time according to their timestamps. Real-time renderers are always associated to a **TimeSource** which represents the temporal progression of fragment processing (see figure 3.6), and applications may use the information provided through the **TimeSource** to provide synchronization with other media processing.

The **MediaSource**s or **MediaRenderer**s aggregated by a **RealTimeSource** or **RealTimeRenderer** are always synchronized to each other and the **TimeSource** supplied by the parent object; this is usually achieved by using the callback mechanism of the time source to drive media processing (see figure 3.7).

Both real-time sources and renderers will usually be subject to *buffering delay* – this does in practice mean that **Fragment**s delivered by a **Source** will carry timestamps in the past, while the application has to make sure that **Fragment**s delivered to a **Renderer** carry timestamps in the future (relative to the respective **TimeSource**s). Both **Source** and **Renderer** interface allow applications to query required delays[12], but ultimately the application has to setup media processing appropriately to incorporate the required delays. If both sources and renderers are real-time this is generally achieved by "offsetting" the **TimeSource** used for rendering appropriately from the source **TimeSource**. This delay is completely under the application's control, the media framework does *never* introduce any processing delays into the media processing pipeline on its own.

# CHAPTER 3. MEDIA PROCESSING FRAMEWORK ARCHITECTURE

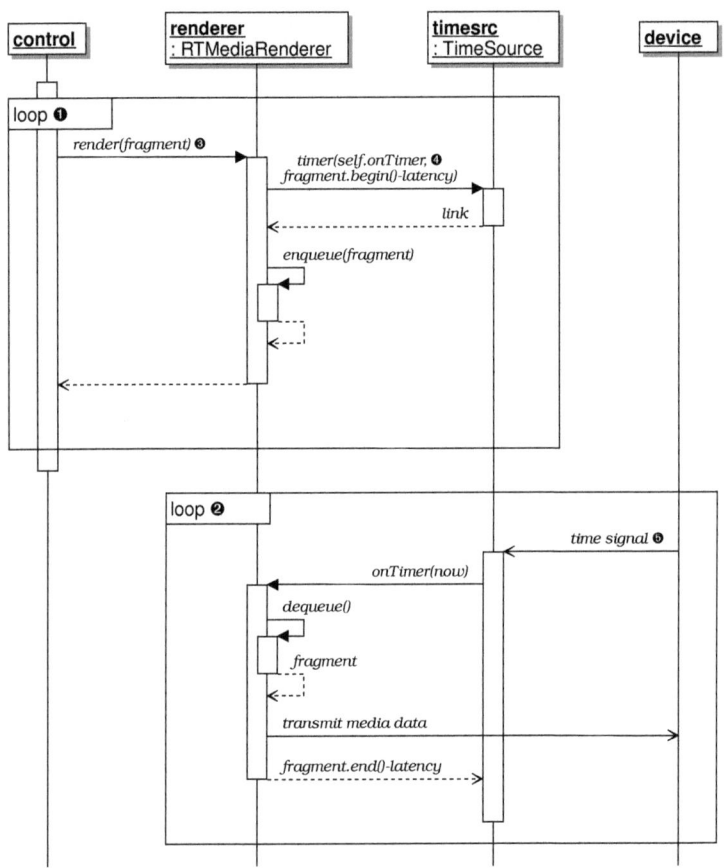

In practice, loops ❶ and ❷ are executed asynchronously and in parallel. The application passes fragments to the renderer ❸ which will schedule operations to process the fragment, taking into account any transfer latency ❹. The time signal ❺ of the target device is then used to drive processing of the media fragments.

Figure 3.7: Interaction of time sources and real-time renderers

## 3.2. CORE ARCHITECTURE

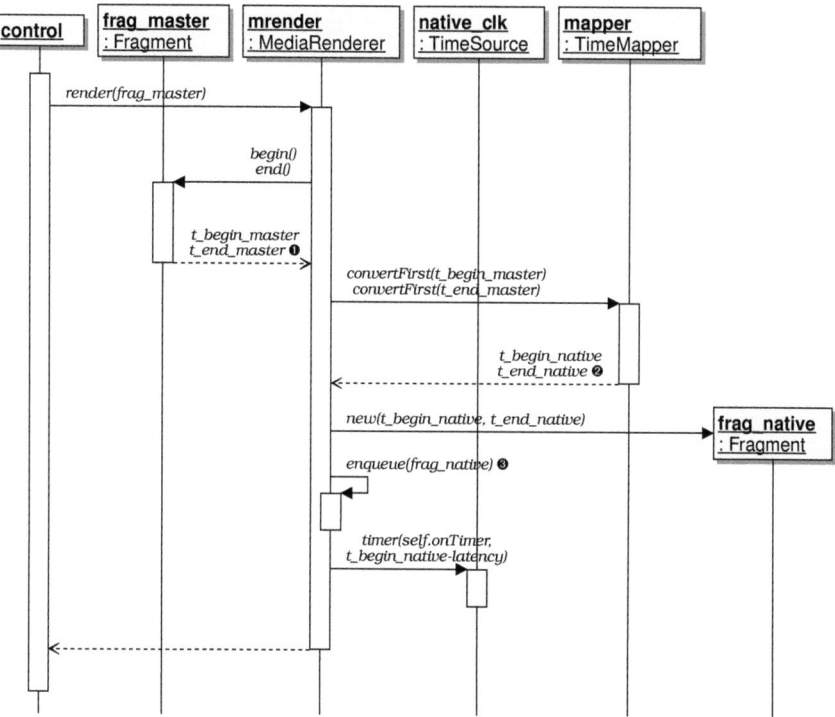

The **MediaRenderer** converts the timestamps of **frag_master** that are relative to the application-chosen master presentation time source ❶ to timestamps relative to the native time source ❷ and creates a new **Fragment** with the translated timestamps ❸. All further processing of the media data is performed using the converted fragments.

Figure 3.8: Synchronization of media processing by mapping time sources

### 3.2.3.5 Synchronized media processing

Real-time rendering is performed at a rate indicated by the **TimeSource** exposed through the interface of **RealTimeRenderer**, but in actual implementations the presentation may be driven in two different ways.

First, the renderer may use the callback mechanism of **TimeSource** to execute code on the host CPU to ensure that a fragment is presented during the correct temporal interval. In this case any source of time information provided by the architecture may be used exchangeably. Second, rendering may be driven by a different entity than the host CPU with its own time source, independent of the host CPU, e.g. a clocked audio DAC. In this case the time source used for presentation cannot be replaced by a different one; this non-replaceable time source will be referred to as "native time source" below.

Applications may need to synchronize media presentation with a given time source (e.g. derived from input devices). In the first case above this can trivially be achieved by exchanging the time source used to deliver callbacks to the renderer driver. In the second case however there is no other way but to use the **TimeMapper** service to map timestamps of the designated master presentation time source to the native time source used by the renderer.

The architecture strives to make implementation differences between different renderer drivers transparent to the application and thus allows to assign a master presentation time source to all **RealTimeRenderer** drivers. Renderer drivers must therefore support to transparently instantiate a **TimeMapper** and convert timestamps, as needed (see figure 3.8).

## 3.3 Media type support

The core media processing library is agnostic with respect to media types such as audio, video or images. It provides abstract support for time-dependent media through the **Fragment** concept on which specific media types such as audio and video can be built. It is explicitly designed in a way that makes definition of new media types easy.

As explained in section 3.1.2, the base abstractions provided by the library treat these media types as time- and space-continuous. These abstractions are sufficiently generic to allow *any* computable image or audio representation conceivable into the architecture.

All media elements support the concept of *processor-private data*: Any **Processor** instance may attach additional data objects to any media element – the attached data is however private to the **Processor** instance that attached the data (i.e. it is inaccessible for all other **Processor**s) – the interpretation of the data is up to each **Processor** itself. This provides a mechanism to "annotate" media ele-

---
[12]For **Sources** the delay can be determined implicitly by querying the reference clock for every fragment received by the application.

3.3. MEDIA TYPE SUPPORT                                          101

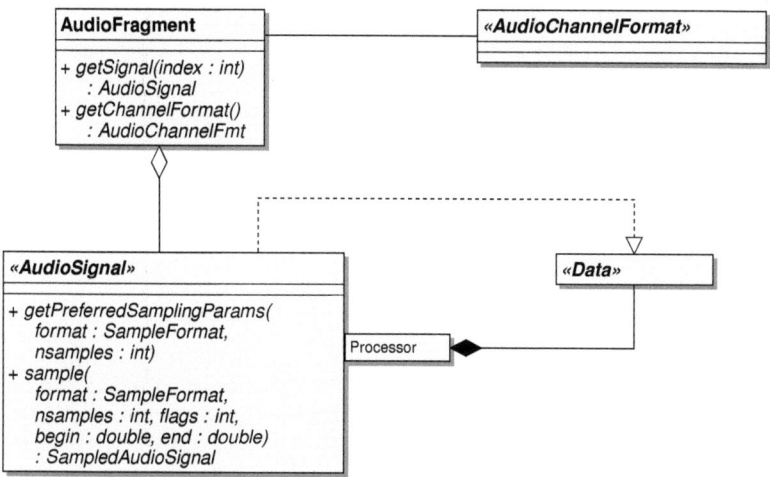

Figure 3.9: Audio representation

ments with additional data, e.g. to cache results of computations performed by a specific processor.

### 3.3.1 Audio

Audio support is based on specializing the **Fragment**, **Source** and **Renderer** concept and the introduction of the **AudioSignal** interface (see figure 3.9).

**AudioSignal** represents an arbitrary (computable) function, mapping (a superset of) the interval $[0; 1)$ to $\mathbb{R}$ (cf. the definition 1 on page 12). It is an abstract interface that defines the **sample** method which allows to sample the function it represents at a set of equidistant points (which is why the underlying function must be computable).

Each **AudioFragment** contains an **AudioChannelFormat** that describes the semantics of all audio channels active during the temporal interval of this fragment, as well as one **AudioSignal** for each channel. The intended semantic is that the signal functions represent the pressure intensities at the spatial positions described by the channel format for the temporal interval represented by the fragment.

#### 3.3.1.1 Audio signals

The **AudioSignal** interface serves as the base class for many other possible representation alternatives:

- **ConstantAudioSignal**: Represents a signal that evaluates to a constant.
- **LinearAudioSignal**: Represents a linear function.
- **FunctionalAudioSignal**: Represents a signal computed from a *functional*[13]. (The purpose of this class is to act as a convenient wrapper to let programmers represent arbitrary computations as **AudioSignal**s.).
- **SampledAudioSignal**: Represents the canonical band-limited interpolation of a time-discretized audio signal sampled at equidistant points (see sections 1.3.2.1 and 1.3.2.2).
- **CompressedAudioSignal**: An audio signal in a compressed representation (see section 3.3.3).

In addition to the "non-algebraic" representations above, the following classes allow to represent an audio signal as the function that would result from the application of an algebraic operator on one or more other given audio signals:

- **SumAudioSignal, ProductAudioSignal**: Sum and product of (at least) two audio signals.
- **ConvolutedAudioSignal**: An audio signal that results from the application of a *convolution filter* to another audio signal.
- **SlicedAudioSignal**: Given a signal function $s$ and $0 \leq t_{begin} < t_{end} \leq 1$, represents the signal function $s'(t) = s((t_{end} - t_{begin}) \cdot t + t_{begin})$. (In other words, the operator "slices" the interval $[t_{begin}; t_{end})$ out of $s$ and "stretches it to $[0; 1)$.)
- **ConcatAudioSignal**: Given two signal functions $s_1$, $s_2$ and a constant $t_0$, represents the signal function

$$s(t) = \begin{cases} s_1(t/t_0) & \text{if } t < t_0 \\ s_2((t-t_0)/(1-t_0)) & \text{if } t \geq t_0 \end{cases}$$

(In other words, the operator temporally "concatenates" the signals $s_1$ and $s_2$.)

- **TransferAudioSignal**: Given a signal function $s$ and a monotonic function $f$, represents $s'(t) = f(s(t))$.

Combining the above alternatives allows to construct *syntax trees* that represent audio signals as complex arithmetic terms[14]. Note that this includes the operators discussed in section 1.2.1 as a subset.

---

[13] The term *functional* here has the meaning of a class that can be called as a function, i.e.:
class SineFunctional {
 public:   double operator() (double x) {return sin(x);}
};
[14]The data structures will in reality actually be directed acyclic graphs since "sub-trees" can be shared. This should be regarded as an "optimization" that a) saves storage space over a strict tree representation and b) simplifies identification of common subexpressions.

## 3.3. MEDIA TYPE SUPPORT

### 3.3.1.2 Sampled audio signals

In practical applications audio is often represented as a sequence of temporally equidistant sample values (cf. section 1.3.2.1). This feature is provided through the **SampledAudioSignal** class. Each object of this class represents $n$ sample values taken at the points $0, 1/n, 2/n, \ldots (n-1)/n$. Longer sequences of sampled audio can be represented as multiple audio fragments, each containing a **SampledAudioSignal**.

While the definition of **SampledAudioSignal** is intuitive and appears innocent, the actual interpretation as band-limited interpolation of the discrete sample data (see section 1.3.2.2) has a number of interesting consequences.

Assume an audio signal given by the function $f : \mathbb{R} \to \mathbb{R}$. Further assume that the signal is band-limited by $n$ (or, in other words the Fourier-transform $\mathcal{F}(f)$ is supported by an interval of (at most) length $n$).

This audio signal will now be represented through a countable number of sample values. Let $f_s$ be the function defined by

$$f_s(t) = \left\{ \begin{array}{l} f(t), t = \frac{k}{n}, k \in \mathbb{Z} \\ 0, \text{else} \end{array} \right\}$$

i.e. the function $f$ sampled at points $k/n$. According to the Whittaker-Shannon interpolation formula the function $f$ can be represented as

$$f = (f_s * \delta) * (t \mapsto \operatorname{sinc}(n \cdot t)) \qquad (3.1)$$

or equivalently:

$$f(t) = \sum_k f_s\left(\frac{k}{n}\right) \operatorname{sinc}\left(n \cdot t - \frac{k}{n}\right) \qquad (3.2)$$

Both $f$ and $f_s$ are conceptually infinite signal functions which – for the purpose of processing within the media framework – have to be split up into smaller fragments. Let a fragment represent the temporal interval $[0;1)$. The audio signal during this interval is given by a function $f' : [0;1) \to \mathbb{R}$, $f'(t) = f(t)$ but a **SampledAudioSignal** object would instead store $f'_s : \{0, 1/n, \ldots (n-1)/n\} \to \mathbb{R}$, $f'_s(t) = f'(t)$ which is $f'$ sampled at points $0, 1/n, 2/n, \ldots (n-1)/n$.

However, as equations (3.1) and (3.2) show, $f'$ *cannot* be reconstructed from $f'_s$ alone, instead this requires sample values from $f_s$ outside the interval $[0;1)$. The signal represented by a **SampledAudioSignal** is thus given from the samples by:

$$(f'_s * \delta) * (t \mapsto \operatorname{sinc}(n \cdot t))$$

which can obviously be non-zero outside $[0;1)$. As a consequence adjacent fragments can not simply be "concatenated" but must instead be summed with appropriate temporal offset to reconstruct the original signal (see figure 3.10).

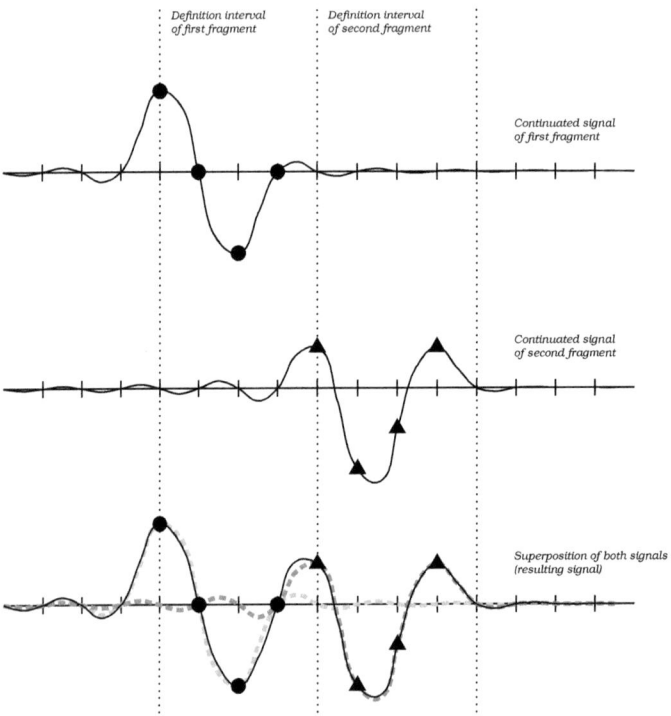

Figure 3.10: Band-limited continuation of **SampledAudioSignal**s

Conceptually, each fragment contains only data corresponding to its own validity interval; it is also possible to reconstruct $f_s$ from these fragments, and each fragment will only contribute data within its own interval. However, in order to properly reconstruct $f$ as a continuous function each fragment has to contribute data *outside* its time interval. (Note that this interpretation results from the assumption that $f$ is properly band-limited.)

Note that **SampledAudioSignal**s can be assumed to be "cross-talk free" unless the signal needs to be resampled – this is generally the case only when

- multiple audio signals with mismatched sampling rate are to be combined (e.g. **SumAudioSignal**)

- a media processor (such as an output device) requires a specific sampling rate different from the one the data currently is represented as

## 3.3. MEDIA TYPE SUPPORT

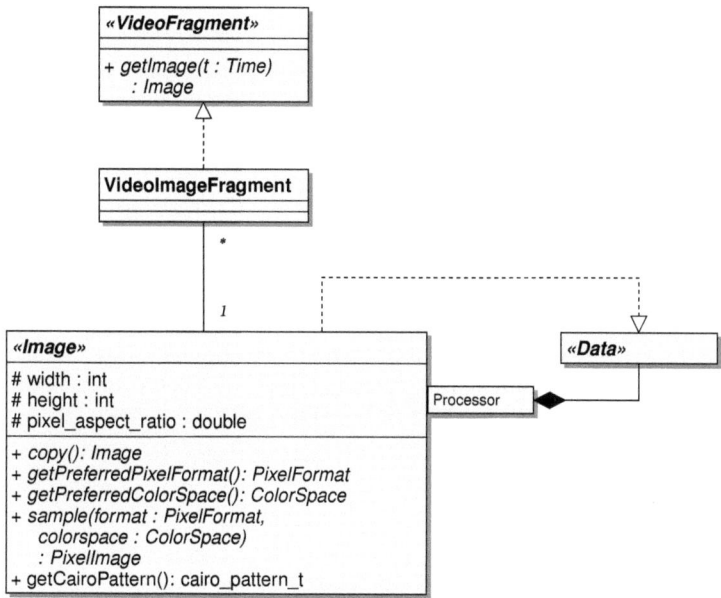

Figure 3.11: Video representation

### 3.3.1.3 Channel formats

The **AudioChannelFormat** object referenced by every fragment provides the intended interpretation of all audio channels within a fragment. Generally, this interpretation is given in the form of "positional information" for each channel.

The information provided by the **AudioChannelFormat** may be purely logical "tags" for each channel (e.g. "left stereo channel", "right stereo channel", "low frequency effects (LFE) channel"). Several predefined **AudioChannelFormat**s provide support for common audio configurations such as mono, stereo or multi-channel surround sound models (e.g. 5.1). But **AudioChannelFormat**s can also provide considerably more complex interpretations for the channels such as a physical acoustic model with spatial coordinates for every channel; in this case the parameters (including position) of each channel need not even be constant over the duration of a fragment.

## 3.3.2 Still images and video

Video support is based on specializing the **Fragment**, **Source** and **Renderer** concept and the introduction of the **Image** interface (see figure 3.11).

Each **VideoFragment** must for each point in its temporal interval be able to supply an image that is supposed to be shown at the corresponding point in time (thus video is conceptually time-continuous, see definition 5 in section 1.1.2.3). It is a common special case that the same image is shown for the full duration of the fragment (see definition 12 in section 1.3.5.1) – this is provided by the specialized class **VideoImageFragment**.

Each **Image** in turn represents an arbitrary (computable) function that assigns a color (using one of the methods outlined in section 1.3.3 to identify colors with value triplets) and an alpha value to each point of the rectangle $[0;w) \times [0;h)$ (see definition 4 in section 1.1.2.2). It is an abstract interface that provides the **sample** method which can sample the image using a selected pattern (see section 3.3.2.2 below).

### 3.3.2.1 Images

Each **Image** has an associated *size* that is given by its **width** and **height**; these are measured in pixel units which are assumed to be small rectangles with their physical width-to-height ratio given by **pixel_aspect_ratio**. Note that this does not necessarily imply that the image is composed from $width \cdot height$ rectangular tiles – conceptually the image might be an infinitely scalable vector graphics, but **width** and **height** may provide a "hint" as to how the image could be sampled appropriately. An **Image** may supply additional hints for rasterization by providing a prefered **PixelFormat** and **ColorSpace** (see 3.3.2.2). The indicated format pair should provide a representation into which the image can be rasterized without losing information due to rounding and/or subsampling.

The following representation alternatives are provided for images:

- **PixelImage**: A rastered image (see section 1.3.4.1)

- **PaintedImage**: An image defined through a chain of compositing operations (see section 1.2.2).

- **CompressedImage**: An image given in a compressed representation, see section 3.3.3

Calling the **sample** method will instantiate a **PixelImage** that represents the result of sampling the image. Although convenient, this method has to create the full rastered data at once which can be quite inefficient. For practical purposes it is preferable to use the related **getRows** method instead, which returns a **RowIterator** object that allows to access the sampled data one sample row at a time (and may therefore be able to avoid creating a full **PixelImage**).

## 3.3. MEDIA TYPE SUPPORT

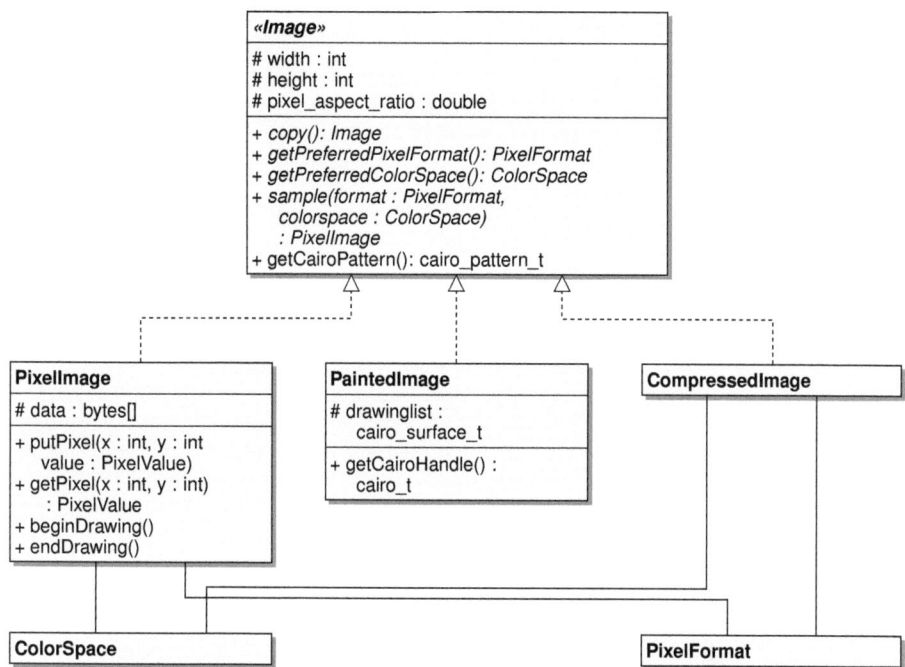

Figure 3.12: Image types

### 3.3.2.2 Pixelformats and color spaces

The **PixelFormat** and **ColorSpace** classes provide a description of how exactly a rastered image is represented as a sequence of bits. **ColorSpace** defines how triplets of values are to be interpreted as colors (see section 1.3.3). Internally, this is done by representing the mapping relationship to CIE 1931 XYZ (section 1.1.2.1, page 14), however **ColorSpace** may also represent pseudo-colorspaces such as "unknown RGB" for which no strictly defined mapping exists[15].

**PixelFormat** defines how the color triplets (and the alpha value) assigned to each pixel are stored in memory, and how many bits of precision are stored for each value. The description supports packed and planar representations, allows channels to be subsampled or omitted, and can also describe "interlaced" images where the full image actually consists of two interleaved fields (see A.5 for a formal description of **PixelFormat**s).

---

[15]In case conversion into another color space is coerced, a simple gamma-corrected color model with primaries matching that of the target color space is assumed.

### 3.3.2.3 Discretized Images

Space-discretized images are represented as **PixelImage** objects; conceptually, they are a two-dimensional array of rectangular pixels and store an associated color value for each pixel. **PixelImage**s serve two main purposes that enable data exchange:

- They allow to introduce images into the architecture for which there is no other representation available.

- They allow to coerce images into a format suitable for further processing external to the media processing framework.

**PixelImage**s permit direct access to the underlying pixel data to support these use cases without the overhead of creating data copies. Since **Image**s are normally treated as immutable objects, the application is expected to modify images only if it can be certain that is has exclusive access. To avoid excessive copying, all **Image**s support the creation of shallow copies through the **copy** method, while **PixelImage**s support "copy-on-write" by bracing modifications to the pixel data with **beginDrawing** and **endDrawing**.

### 3.3.2.4 Painted images

The **PaintedImage** class allows to specify compositing operations to generate a desired image. The model follows the "painter's" algorithm: Starting with a blank image, other image sources and patterns (such as uniform colors or gradients) are combined layer by layer with the previous image contents. The source images and patterns may be geometrically distorted, clipped by path-delimited masks and composited with the lower layers using a variety of operators (see section 1.2.2). **PaintedImage** follows the retained-mode processing model: All operations (including source images and patterns used) are recorded and can later be replayed on a target image[16]. Conceptually, **PaintedImage**s are vector images and are thus space-continuous.

The drawing facilities are exposed to applications through the `cairo` imaging API (see 2.2.2.1) – this choice was made since it is a familiar and sufficiently abstract API, even though the integration turned out to be technically difficult. To provide the bindings to `cairo`, every **PaintedImage** exposes a `cairo_t` drawing context through which operations can be issued. Additionally, every **Image** exports a `cairo_pattern_t` handle to make it usable as a source pattern for compositing.

**PaintedImage**s are therefore created incrementally by the application and can recursively reuse other (painted) images. This puts several responsibilities to the creator of a painted image: First, incremental painting must be finished before the image is handed over to any other processing stage (as it is generally assumed

---

[16]The target image may in fact itself be a **PaintedImage**.

## 3.3. MEDIA TYPE SUPPORT

that the objects are immutable). Second, the application must not create any dependency loops in the graph that represents the participating images on their source patterns.

### 3.3.3 Compressed media

Section 1.4 introduced the concept of "compressed" representations for audio signals and (sequences of) images. Specifically, definitions 13 (page 42) and 14 (page 50) introduced the concept of *temporally local* compressed representations. libmedia supports all types of compressed audio signals and images that satisfy the constraints[17] set out in the definitions through the following set of classes:

- **CompressedAudioSequence** and **CompressedImageSequence** store all information common to a "sequence" of multiple compressed media elements; usually these are common encoding parameters as well as a description of the compression format to be used. This corresponds to the *initialization data* from definitions 13 and 14.

- **CompressedFrame** acts as container for the storage of compressed media data. Like **AudioSignal**s and **Image**s it also supports the "processor-private data" concept.

- **CompressedAudioSignal** references one or more **CompressedFrame** objects as well as one **CompressedAudioSequence** object. It represents the audio samples of one channel that can be generated from the compressed representation.

- **CompressedImage** references one **CompressedFrame** object, one **CompressedImageSequence** object and zero or more **Image** objects used as reference images. It represents the image that can be generated from the compressed representation, using the given reference images.

Note that the above objects merely *represent* the data without being able to actually *process* it – while **CompressedAudioSignal** and **CompressedImage** must support the **sample** method, they delegate to a decompressor instance as discussed below.

Handling of compressed media data consists of several layers in this architecture that perform different levels of services and analysis of the data. This is a significant deviation from other architectures (which mostly put all of the functionality separated here into a single layer). It is however necessary in this architecture to maintain the retained-mode processing semantics (i.e. no decompression operation is performed until "late" in the processing stage) and maintain the logical independence of individual frames (providing the application with random-access semantics).

---
[17]Note that the constraints are in fact not too severe, as already stated all formats in practical use fit into these restrictions.

### 3.3.3.1 Parsers

The first layer responsible for handling of compressed data are *Parsers*. Their role is to inspect the bitstream, partition it into indivdual frames, and extract required meta-data into a generic form usable by the framework. Parsers do not perform decompression – they only interpret the given data to the extent it is necessary to understand its structure.

Parsers are inextricably linked to a specific compressed representation format, as such they are realized as replaceable components. The amount of parsing required however largely depends on the container format the compressed data is stored in (as the container may or may not already provide some of the meta-data), thus the parsers are not generic components but sub-components of file format handlers (see section 3.5.2). No interface is prescribed by the architecture for the communication between parsers and file format handlers as it is both difficult and needless – defining the interface is the responsibility of the file format implementor. Parsers therefore only exist as concepts.

For audio, the parser's role is to identify individual audio frames in a bitstream (unless the container format already separates frames), identify the number of samples represented within one frame as well as the relationship (e.g. overlapping) to adjacent frames. The parser must also extract meta-data such as number of channels and their meanings, whether this is given in a stream header or each individual frame. It is then the parser's responsibility to instantiate **CompressedAudioSequence**, **CompressedAudioSignal** and **CompressedFrame** objects that properly represent the meta-data to the rest of the framework.

The role of image and video parsers is quite similar. Features such as width, height, aspect ratio, color model and pixel format (that describes the channel subsampling) must be extracted from the compressed frame data. Video parsers must additionally understand the coding types (intra or non-intra) and the temporal coding relationships of the individual frames – sometimes the container format allows to specify these explicitly (e.g. the QuickTime file format), sometimes they are implicitly given through the ordering of frames (cf. figure 1.15, page 47). The parser is then responsible for instantiating **CompressedImageSequence**, **CompressedImage** and **CompressedFrame** objects.

### 3.3.3.2 Decompressors

The decompressors provide the lowermost layer of media processing offered by the framework and thus must finally break with the retained-mode processing paradigm. The architecture provides the **AudioDecompressor** and **ImageDecompressor** interfaces that represent decompressor instances that process frames using a specific parameter set (that is normally stored in **CompressedAudioSequence** and **CompressedImageSequence** sequence objects). They are instantiated by **AudioCodec**s and **ImageCodec**s for which a name mapping registry is provided to allow extensibility.

## 3.3. MEDIA TYPE SUPPORT

The decompressor instances rely on the work performed by the parsers to decompose and extract meta-data: They are stateless, processing the same data/meta-data must produce the same result. This is in stark contrast to QuickTime or DirectShow, where e.g. image decompressors are stateful and responsible for identifying and preserving frames that are used as references in the future.

Essentially, **AudioDecompressor**s and **ImageDecompressor**s perform the operations described before defenition 13 (page 42) and 14 (page 50): They transform the compressed data into a sequence of audio samples or a rastered image representation. While **AudioDecompressor**s generate the sample data in a single operation, **ImageDecompressor**s provide an iterator-based interface that allows to incrementally process the image as it is being generated.

The **AudioDecompressor** interface is designed such that it takes possible frame overlapping into account – as a consequence it may receive the same frame multiple times as it forms part of a larger group. To avoid repeating computations already performed on the frame data, it may use the processor-private data mechanism to attach generated data for future reference.

### 3.3.3.3 Compressors and sequence managers

Transformation of audio and image data into a compressed representation is facilitated through the the **AudioCompressor** and **ImageCompressor** interfaces (instances of which are created in the same way as decompressors). Normally, applications will never explicitly instantiate and use these objects and rely on renderer drivers to implicitly use the compressors if required, see section 3.4.3.

Instead, they are generally called by *sequence managers* that are the inverse of parser objects: sub-components of container objects that are familiar with both the structure of the container format as well as the compressed media representation format. They are responsible for controlling encoding of frames and enforcing restrictions (e.g. permissible coding types and their temporal relationship) imposed by the combination of media and container format.

In principle, the compressor interface is a straight reversal of the decompressor interface (i.e. **ImageCompressor**s turn a given image into a bitstream, using specified reference images). One important difference however is that the compressors may return objects that are not exactly coded as requested – **ImageCompressor** may produce an intra-coded frame even though a predicted frame was requested (because prediction turned out to be not useful), in a similar way **AudioCompressor**s may code smaller frames than requested (if the format allows variable-sized frames). This in turn requires the sequence managers to adapt a "back-off and retry" approach to coding as the compressor's decisions may from time to time disrupt the planned sequence structure.

### 3.3.4 User-defined representation types

The architecture defines abstract interfaces **Image** and **AudioSignal** that serve as representation for the respective media element types. Several pre-defined implementations of these interfaces where presented in this chapter, one of those a quite generic "compressed" representation. The architecture is of course extensible: Any application or module may create "user-defined" representation types by simply inheriting from these two interfaces.

Extensions may also define new media processing operations. In keeping with the retained-mode processing concept, these operations should just create new media elements from the given ones. Where the pre-defined representation alternatives and operations encapsulated within are insufficient, newly defined implementations of **Image** or **AudioSignal** may be used in this place: For example, a function **spectacular_effect** for providing a particular effect on one (or more) images would take one (or more) **Image** objects as input parameter and produce a **SpectacularEffectImage** as output.

**SpectacularEffectImage** must at a minimum implement the **sample** operation which computes the effect – it may in places be is used as a "safe fallback" to convert the data into a format that the rest of the framework understands. It is conceivable that the architecture provides more efficient special-case handling for the "known" representations provided above – which may certainly be desirable for performance improvements. While it would at first glance appear that this would reduce user-defined types to "second class citizens", the *renderer* concept – which will be discussed in section 3.4.3 – is sufficiently generic to be extensible as well. An illustrative example for this extensibility will be given for the **X11VideoRenderer** in section 4.3.3.

## 3.4 Processing

For the most part media processing using the framework consists of the following steps:

- Obtaining fragments from **Source**s/**MediaSource**s

- Possibly applying some transformations on the media

- Handing the data over to **Renderer**s/**MediaRenderer**s for interpretation

Of course, applications have other options as they can "synthesize" the fragments instead of using the **Source**s, or they can interpret the data itself without the use of a **Renderer** driver.

## 3.4. PROCESSING

### 3.4.1 Compositing

Applications that wish to apply any transformations on media data must follow one simple but important rule: All data elements (**Fragments**, **AudioSignals** and **Images**) must be treated as *immutable* – the only exception is of course the creator of the element who may modify it to bring it into its final state up to the point where it is passed on.

This means that generally data elements may not actually be *modified*, but it is permissible to construct new data elements using given data elements as inputs. This is acceptable as there is no data copying overhead due to the retained-mode processing paradigm.

It should be obvious from the description in sections 3.3.1.1 and 3.3.2.4 how new data elements can be composited from old ones. The library offers a few convenience functions that wrap construction of the composited media elements.

For generic time-based media, fragments support creation of time-shifted fragments of identical media content (essentially applying the *prototype* pattern).

For audio, helpers like **sum**, **scale** or **mix** just create new chains of **AudioSignal** objects that represent the arithmetic operations required – these are trivial as the temporal duration of the output equals the duration of the inputs. The situation is more complicated for FIR filters – while each fragment can be processed individually, the convolution operation slightly "widens" the temporal duration of the output (by the width of the kernel), leading to overlap. For this purpose, a small (stateful!) helper can take care of distributing the overlap to the adjacent fragments.

For images, applications need to create a **PaintedImage** and incrementally apply drawing operations until it has the desired appearance. For video, a helper function wraps extraction of images out of a **VideoFragment** and repacks the data into fragments.

While applications can always use the **sample** methods to extract discretized representations out of any media element, they should refrain from doing so unless necessary as this obviously negates the benefits of retained-mode processing.

### 3.4.2 Capture

Media data acquisition devices are represented through the **CaptureDevice** abstraction. It provides facilities to enumerate the different channels available through this device (e.g. for video/audio capture) and exposes controls that allow applications to setup capture parameters to be used for a session. `libmedia` does not provide device enumeration capabilities or any **CaptureDevice** implementations – these are system-specific services provided by separate modules.

The main purpose of **CaptureDevice**s is to instantiate specialized objects implementing the **Source**, **AudioSource** and **VideoSource** interfaces. In addition to the objects providing the media data, capture devices also supply **TickSource**

114    CHAPTER 3. MEDIA PROCESSING FRAMEWORK ARCHITECTURE

or **TimeSource** objects that allow applications to synchronize other processing with the capture rate. The source objects produce correctly timestamped fragments containing an **AudioSignal** or **Image**, using any of the representation alternatives discussed above that best fits the underlying device (typically this will either be **PCMAudioSignal/PixelImage** objects or **CompressedAudioSignal/CompressedImage** objects) – as usual, further applying processing operations may cause implicit conversions into a different format.

Capture sources are generally real-time capable (see section 3.2.3.4) and thus applications have the option of using synchronous blocking I/O (e.g. if the application prefers to spawn dedicated processing threads for each source) or asynchronous non-blocking I/O in conjunction with readiness callbacks (e.g. if the application prefers an event-driven model), refer to section 3.2.2.

### 3.4.3  Rendering concept

Since media processing is centered around the idea of retained-mode processing, the media elements received by **MediaRenderer**s are normally just abstract descriptions how the data element could be computed. Obviously these descriptions must ultimately be interpreted and adapted to the specific target, be it a playback device or storage into a file.

While renderer drivers are ultimately free to process the data as they please, there are several tasks common to almost any driver, so that it makes sense to provide architectural support for these tasks.

#### 3.4.3.1  Channel format converter services

Audio renderers may receive audio fragments using a **ChannelFormat** that does not match that of the render target. The renderer must thus be able to convert to a desired channel layout. To avoid code duplication, the architecture provides the *channel format converter service*: It allows instantiation of converters to transform audio from one format into another. The converters will receive fragments in their designated input format and produce fragments in their designated output format.

The converters themselves have access to the full range of processing capabilities (and may even use other auxiliary converters themselves). While they may process the input data using any method and represent the output data in any way they wish, they *should* stay within the architecture and express their operations using the abstract operations presented in section 3.3.1.1. This section will show what kind of format conversions can be supported using these operations.

Let $f_1$ through $f_n$ denote the signals associated with the $n$ input channels and $g_1$ through $g_m$ denote the signals associated with the $m$ output channels. Through time-dilation of input signals as well as elementary arithmetic operation the following transform can be specified using the mechanisms from section 3.3.1.1:

## 3.4. PROCESSING

$$\begin{pmatrix} g_1(t) \\ g_2(t) \\ \vdots \\ g_m(t) \end{pmatrix} = \begin{pmatrix} a_{11}(t) & a_{12}(t) & \cdots & a_{1k}(t) \\ a_{21}(t) & a_{22}(t) & \cdots & a_{2k}(t) \\ \vdots & \vdots & \ddots & \vdots \\ a_{m1}(t) & a_{m2}(t) & \cdots & a_{mk}(t) \end{pmatrix} \begin{pmatrix} f_{n_1}(\alpha_1 t + \delta_1) \\ f_{n_2}(\alpha_2 t + \delta_2) \\ \vdots \\ f_{n_k}(\alpha_k t + \delta_k) \end{pmatrix} \quad (3.3)$$

(We assume that the functions $a_{mk}$ are independent from all functions $f_n$. This representation is already sufficient to cover the most common case of conversion between different surround-sound formats, e.g. a downmix transformation from 5.1 to stereo can be achieved through:

$$\begin{pmatrix} g_{left}(t) \\ g_{right}(t) \end{pmatrix} = \frac{1}{1 + clev + slev} \begin{pmatrix} clev & 1 & 0 & slev & 0 \\ clev & 0 & 1 & 0 & slev \end{pmatrix} \begin{pmatrix} f_{center}(t) \\ f_{left}(t) \\ f_{right}(t) \\ f_{left\_s}(t) \\ f_{right\_s}(t) \end{pmatrix}$$

with $clev = 1/\sqrt{2}$ and $slev = 1$, or determined by further downmixing hints in the **AudioChannelFormat** description.

However, the transformation shown in equation (3.3) is also sufficient to express simple rendering of positional point audio sources in an empty space. To illustrate this consider a set of point audio sources, each emitting audio signals described by the functions $f_1(t)$ through $f_n(t)$, and each of which may be in motion relative to the observer. The goal is to simulate this acoustic environment to an observer given a set of $m$ static speakers – this means that the signals $g_1(t)$ through $g_m(t)$ to be emitted by each speaker must be calculated.

The signals $g_j(t)$ can then according to Huygen's principle be calculated as

$$\begin{pmatrix} g_1(t) \\ g_2(t) \\ \vdots \\ g_m(t) \end{pmatrix} = \begin{pmatrix} a_{11}(t) & a_{12}(t) & \cdots & a_{1k}(t) \\ a_{21}(t) & a_{22}(t) & \cdots & a_{2k}(t) \\ \vdots & \vdots & \ddots & \vdots \\ a_{m1}(t) & a_{m2}(t) & \cdots & a_{mk}(t) \end{pmatrix} \begin{pmatrix} f_{n_1}(\gamma_1(t)) \\ f_{n_2}(\gamma_2(t)) \\ \vdots \\ f_{n_k}(\gamma_k(t)) \end{pmatrix} \quad (3.4)$$

In this model the functions $\gamma_j(t)$ represent the time-dilation due to wave propagation latency while the coefficients $a_{ij}(t)$ correspond to energy loss due to propagation into space. (Finding a suitable transformation however is still a non-trivial task).

Since the functions $\gamma_j(t)$ represent the time dilation, their value depends purely on the relative movement with respect to the observer. Assuming that the motion is "slow" compared to the speed of propagation of sound, the functions can very well be approximated with piecewise linear functions, with precision well below the human perception threshold. Thus (3.3) provides a suitable approximation to (3.4).

It should be noted that (3.4) does in principle allow to incorporate environmental effects such as reflections and reverberations – however, the representation is not particularly well-suited to actually calculate these effects[18].

### 3.4.3.2 Optimization

A simplistic approach were to use the **sample** methods on audio and image objects to transform them into a representation that is relatively easy to understand and process. However, this loses out on some of the more interesting opportunities offered by the conceptual approach taken in this architecture for optimization: Essentially, the renderer drivers act as "compilers" that understand the operations specified in the input data, analyze them, apply optimization transformations, and translate them for the specific target.

While several standard optimization techniques (such as common subexpression elimination, dead code elimination) are immediately applicable, the interesting twist is that optimization must often be performed with *incomplete information*: The renderer's look into the future is limited by the amount of media data that the application supplies ahead of time (this has the most impact on real-time rendering). This means that optimization is sometimes speculative, based on heuristics, and that future data can invalidate these assumptions. The renderer driver must be prepared for that.

The following sections will explore some of the possible optimizing transformations.

**Scheduling**

Real-time renderer drivers must ensure that all required transformations are performed in a timely fashion – knowing the required total processing time of each data element (typically from observation of the past, but sophisticated metrics can be realized here), the renderer driver can compensate for the processing latency.

This also includes smoothing out processing spikes – in image sequences using bi-directional temporal prediction there are times when an image to be displayed is already fully decompressed (because it was used as a reference frame). The renderer driver can anticipate the next image to be decompressed and schedule this operation into the idle time.

The same mechanism can also be used to schedule frame drops in case the processing load surpasses a set threshold – preferably compute-intensive frames that are not referenced any further should be dropped, renderer drivers are equipped with sufficient information to make an informed decision.

**Inverse transformations**

One of the most obvious optimizations is to cancel out chained inverse operations. While this is usually applied in the context of arithmetic transformations

---

[18]Reverberations can be interpreted as multiple echoes and calculated as sum of time-dilated and attenuated copies of the original signal, however in practice this calculation is performed using FIR filters.

## 3.4. PROCESSING

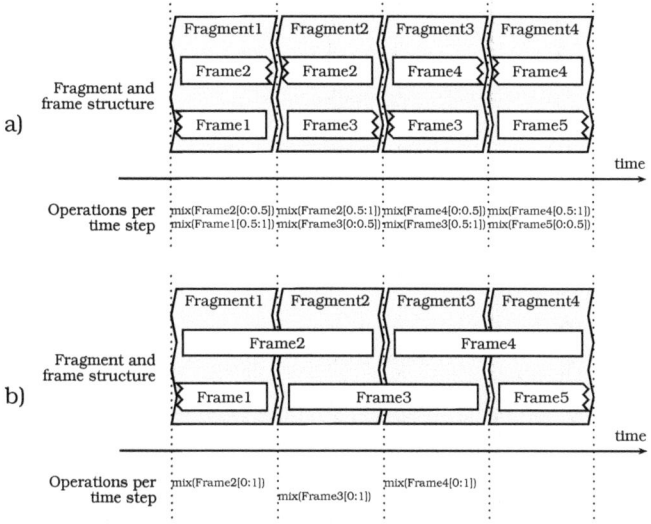

a) Structure auf **AudioFragments** as they are received by the renderer. Due to overlapping of audio frames, the signal contained in each fragment is a mix of the signals obtained from two halves of two audio frames. Executed naively, this would mix in two halves of two audio frames at each time step.
b) Merging of audio frames split over multiple **AudioFragments** allows to process each frame as a whole. Note that the fifth frame is not scheduled for mixing yet, as the renderer speculates that the next fragment will contain the second half of the frame.

Figure 3.13: Compound temporal audio optimizations

to simplify expressions, the operations expressed through the different media element representations are quite high-level.

In its simplest form, this optimization allows e.g. to remove pairs of subsequent compression/decompression operations using the same underlying format and same set of parameters. This optimization alone is already useful as it is quite common in media processing that only a few frames are modified (e.g. for scene transitions, addition of subtitles) while the majority of the frames is left unaltered. More sophisticated options include e.g. reusing common features of source and destination format to allow "partial" decompression and compression (see section 6.3.1 for a discussion).

**Compound temporal audio optimizations**

Audio signals spanning a long temporal period require architecturally to be split up into smaller **AudioFragments** for incremental processing. While it is conceptually desirable to make these fragments "cross-talk free" (e.g. by structuring fragments such that each fragment contains exactly the audio samples contained in one compressed frame when the format is such that frames do not overlap), it is not always possible to do so as related data may be split into multiple adjacent frames.

The "naive" approach to sampling – to sample every fragment independently – might therefore cause underlying signals to be sampled twice if they happen to be split across frame boundaries, which however is inevitable to achieve overlapping. This can be mended by "fusing" frames split over multiple adjacent fragments to process them as a whole (see figure 3.13)[19].

This type of optimization requires speculation as the optimizer may have to be sufficiently "patient" to receive enough fragments to perform the fusing operation, as well as *common subexpression* analysis of the syntax trees representing the operations in each fragment.

### JIT code generation

In some cases the renderer may resort to "just-in-time" code generation for specific tasks: The architecture is fundamentally centered on formally annotating and describing the format a specific data element is represented in and less in providing components that handle the formats. When converting between two different representations, it is generally preferable to use a "direct" transformation that does not use one or more intermediate formats. This is beneficial for both efficiency and accuracy as it helps minimize number of steps and e.g. rounding errors. But since supporting $n$ formats requires $n \times (n-1)$ conversion routines, this can quickly become infeasible.

If the formats are highly structured and quite regular, it is feasible to generate code "just in time" to perform the required conversion. Currently this strategy is utilized for transformations between different color models and pixel formats, see appendix A.5.

### State caching

The fragments processed by the renderer driver are conceptually independent and contain enough information to be processed "stand-alone", so no explicit state is carried over from processing one fragment to the other (unlike filtergraph architectures where there is explicit *hard* state expressed in the processing pipeline).

However, many of the transformation operations do certainly benefit from persistent state (such as decompressor instances that can be reused), so renderer drivers will generally want to cache certain state information (and thus create a certain amount of *soft state*). Renderer-drivers can utilize the *processor-private data* concept (see section 3.3) to retain state, e.g. to attach an **ImageDecompressor** object to a **CompressedImageSequence**.

### Specialized media element type handling

All the **AudioSignal** and **Image** media elements provide the **sample** method to coerce conversion of the data into a sampled representation that is easily understood. This is however far from optimal as the renderer may perform con-

---

[19]Note that this transformation essentially splits the loop iterating over the output samples into two loops, so this optimization is comparable to loop fission in compiler technology.

siderably better if it understands the specific more specialized representations described in sections 3.3.1 and 3.3.2.

Renderer drivers can certainly hard-code handling of specific media type representations, however this would reduce user-defined media types (see section 3.3.4) to second class citizens. Instead, renderers should preferably adopt a dynamic binding scheme using handler class instances for processing of individual media representation types (see also the discussion in section 4.3.3). It is however not useful to define a unified component interface for these handlers as their interface will be very much tied to the specific requirements of the renderer class. Instead, each renderer driver is expected to define its own interfaces for media element handlers as well as corresponding component registries.

This approach to media handler essentially requires a "double dispatch" mechanism (which is generally is not supported as a first-class construct in object-oriented languages): The concrete operation to be performed depends on the types of *two* involved objects, instead of the usual one type in case of virtual method calls [27] [4]. This architecture however takes the concept even one step further in that the handlers need not be defined at compile time of any of the participating objects but may be provided as dynamically loadable components.

**Communication**

Individual renderer drivers may delegate part or all of the processing to other nodes, such as specialized hardware or remote display systems (cf. section 4.3). The renderer must in this case transmit required data to the remote location. In addition to the above optimizations that basically strive to minimize the computational effort required to perform a desired operation, the renderer driver must also take the cost of *communication* into account.

The renderer may often face choices whether to minimize communication at the expense of duplicated processing, or trade in reduced computational effort for more communication. Consider for example temporally compressed images: Reconstructing a sampled image representation in a remote location may require to transmit the compressed data of the image in question *plus* any required reference images. Alternatively, the renderer may also decide to perform all required steps locally and transmit the uncompressed image instead. Depending on the driver's knowledge of the characteristics of the underlying communication channel, it may in certain situations prefer either of the above two approaches. Note that the meta-data provided by the media framework (in the form of dependency chains between the media elements in compressed representation) is crucial for the renderer to make an informed decision.

## 3.5 Documents

The architecture provides the **Document** interface as an abstraction for stored media. Usually, the storage is random-accessible and may allow mixed read/write access. Usually, the underlying storage is realized as a file using

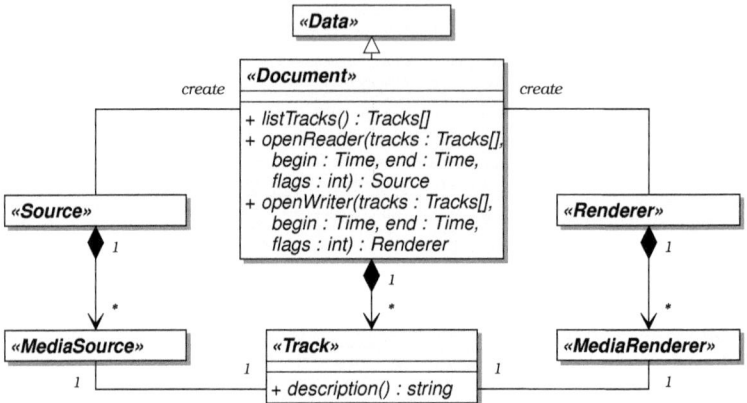

Figure 3.14: Document and accessor objects

a specific container format for on-disk representation, but storages using multiple files or network services are permissible as well – they can also represent pure in-memory "scratch" documents.

Conceptually, the **Document** interface is comparable to QuickTime **Movie**s (see section 2.1.1.2). However, unlike QuickTime this architecture strictly distinguishes between the stateless *document* object itself, and stateful *accessors* used to store and retrieve data. Figure 3.14 shows the relationship of the classes involved.

A **Document** consists of multiple individual **Track**s of media data. Each track must be of one elementary type of time-dependent media (e.g. "video" or "audio", though other types of media can be defined as well). **Documents** provide interfaces for managing (browsing, adding) multiple tracks. As with sources and renderers no interpretation of the relationship of the individual tracks is enforced by the architecture. This must be inferred from additional context; objects implementing the **Document** interface may or may not provide this additional information.

### 3.5.1 Accessors

**Document** objects can instantiate **DocumentReader** and **DocumentWriter** objects (which inherit from **Source** and **Renderer** respectively) that act as accessor objects to the document. Multiple accessors may be instantiated concurrently, however implementations can exclude read/write or multiple writer concurrency if this is not feasible due to the underlying format. Depending on the underlying storage some restrictions may however be imposed on **Document** implementations – applications must therefore query the capabilities of given documents and choose appropriate document types that provide the operations required for their purpose.

Accessor objects are instantiated to start reading or writing a at specific point on the **Document**'s time scale – the mechanism thus allows random-access to the data (though seeking may be imprecise depending on the underlying format). The design assumption is that true frame-level random access is rarely required by applications, as most use-cases involve reading at least short sequences before switching to a new position in a document. The accessor concept is also beneficial from a technical point of view, as extraction of meta-data for identification of coding dependencies usually requires several frames of context information, which the stateful accessors can provide. (Contrast this with QuickTime's frame-level random access concept, section 2.1.1.2).

### 3.5.2 Container file formats

The most prevalent type of multimedia storage is that of a container file. The core architecture has no support for specific file types, but defines a registry for different **FileFormat** components which an application may choose from. To simplify automatic recognition of given file types, it relies on the operating environment's MIME-type facilities for identification, and provides the concept of **MIMEHandler** components (with an assorted registry mapping the system-supplied type names to components) that can instantiate the correct **Document** object class as handler for the particular file format.

Individual file format implementations will normally supply several sub-registries to map symbolic names to handler objects. Typical handler objects are:

- *Track handlers* responsible for instantiation of **Track** information objects as well as track accessor objects

- *Parsers* (see section 3.3.3.1) responsible for meta-data extraction from the coded media data

- *Sequence managers* (see section 3.3.3.2) responsible for controlling generation of compressed media

No generic interfaces for these objects are defined by the architecture, it is up to the **Document** implementor to define interfaces suitable for the underlying container file format.

In addition to these "forward-mapping" registries, the file format handler may also define "reverse-mapping" registries that can e.g. translate **AudioCodec** and **ImageCodec** objects to handlers and the symbolic names used for identification within the file.

# Chapter 4

# Cooperation with the X Window System

The media framework presented in the preceding chapter treats input and output "devices" as replaceable components and therefore is not tied to any particular implementation. Instead, drivers are required (and could be written) for each particular output.

The framework provides good support for "dumb" output devices that can display/playback sampled image and audio data (i.e. a framebuffer-like device, or a simple DAC) as it has built-in methods to convert media data into a sampled representation. However, what sets it apart from other implementations is the ability to support "intelligent" devices that can *dynamically* delegate processing steps to these devices on a case-by-case basis.

This chapter introduces a particular intelligent output device – an extended X Window System – and shows how its processing capabilities can be used. The extensions honor the basic X design principles and logically extend the system from a network graphic and window system to a network multimedia system. The corresponding **Renderer** driver (also described in this chapter) allows application program authors to easily write network-transparent multimedia applications.

**Terminology and UML Notation**. The terms *client* and *server* are always used with the meaning they have in the X Window System (even if referring to other networked windows systems, existing or purely hypothetical): The *client* is the application containing all logic, while the *server* provides display and user interaction capabilites and acts on behalf of the client.

The architecture of the X Window System is fairly object-centric. The terms commonly used in the context of X ("resource", "resource type/class") map reasonably well to the more commonly used terms of object-oriented architectures ("object", "class") and the latter terminology will be used throughout.

Moreover most X requests can fairly well be understood as "methods" of the first class ("resource") involved in the request, and most UML diagrams in this chapter must be read with this in mind. This means that

```
XDrawLine(display, window, ...)
```

becomes
```
window.DrawLine(...)
```
(with `display` implied). The mapping between X requests and class methods resulting from this transformation should be self-explanatory by the name of the method.

The X Window System is designed with asynchronous processing in mind. Therefore, all messages exchanged between client and server (requests and events) are shown as asynchronous messages in the UML diagrams between client- and server-side objects. The message reception and dispatching logic is always omitted from the diagrams as showing the full call chain would only clutter these diagrams.

Actions performed by an object as reaction to a message are shown in the diagrams *as if* the message had been a synchronous procedure call. In practice, messages would of course be batched and processing delayed, this simplification is made in the interest of readability.

Where a diagram involves both classes within the X client and the X server (this will usually be sequence diagrams) they are visually separated by a thick dashed line with labels on both sides indicating the execution context. Classes corresponding to existing X infrastructure (as opposed to the new services introduced in this chapter) will be drawn in diagrams using a hatched gray background (as opposed to uniform light gray) for easy optical disambiguation.

## 4.1 Media processing extensions

In its present form, the X Window System provides little support for multimedia applications. With regards to multimedia applications it must be treated as a "framebuffer-like" device that is just capable of copying an application-generated pixmap to the visible area of the video memory. It does not support audio at all which forces applications to use a secondary audio system.

Technically, these limitations do not pose much of a problem *as long as* the multimedia application in question and the corresponding X display server reside on the same physical machine. In this case, both can communicate using fast inter-process communication mechanisms; they can be considered connected by a communication channel with virtually zero communication latency and virtually infinite communication bandwidth[1]. This means that

- Intra-stream synchronization can (and must) be performed by the application itself; for video it can be achieved by issuing requests to display images to the X server at the correct points in time – the application can be reasonably confident that they will be executed in a timely fashion.

---

[1] Although communication latency and throughput limitations do of course exist they are practically irrelevant for multimedia applications because: 1. IPC latency including operating system scheduling is far below the human perception threshold and 2. IPC throughput exceeds the data rate required to transmit even uncompressed media by orders of magnitudes.

## 4.1. MEDIA PROCESSING EXTENSIONS 125

- Inter-stream synchronization can (and must) also be performed by the application itself; assuming that the application can control and knows the latency of any secondary system used (e.g. for audio) it can again trivially achieve inter-stream synchronicity by timing its requests to the X Window System appropriately.

- Data transfer is easy because it requires (at most) a few memory-to-memory copies.

On the other hand these issues become problematic if the communication channel *does* have non-neglegible latency and *does* have bandwidth limitations of practical relevance, which is of course the case for most network-based communication. Since the X Window System itself works well in networked scenarios, the limitation of multimedia applications to non-networked scenarios becomes quite dissatisfactory.

A simplistic approach would be to construct a multimedia playback system inside (or alongside) the X Window System that autonomously receives, processes and presents multimedia streams[2]. However, this "solution" violates almost every software design principle (of the X Window System in particular) and results in a system that is much more limited in usefulness and less versatile than the approach taken here.

Putting aside for the moment the decision whether multimedia playback facilities properly belong *inside* or *alongside* the X Window System, the following requirements are mandatory for any distributed multimedia presentation system:

- For intra-stream synchronization the *server(s)* have to be able to perform operations (e.g. display images) at precisely defined points in time; since the communication channel must be assumed to have unknown (and unpredictably varying) latency, reliance on timely delivery of commands for individual images or audio samples must be avoided.

- For inter-stream synchronization the *server(s)* must support synchronization between multiple media streams that form a single multimedia presentation; again due to communication latency the system cannot rely on client commands arriving at the server in a timely fashion to ensure synchronicity.

- The system must support transfer of compressed media data because limited throughput (compared to IPC) is of concern for network communications.

Inter-stream synchronization turns out to be complicated (or even impossible) if the client application has to coordinate separate services for multiple media types (i.e. a separate audio and video server process). The server implementor has to

---
[2]like XMovie [36] which provides a pass-through interface for compressed video data

make sure that both services execute synchronously and has to provide a means for the client to express the synchronicity requirements in the command set for the individual servers. The client implementor has to communicate with two separate services, possibly using different protocols. In short, from a systems point of view there is no reason to separate the services to begin with.

This leaves the options of *complementing* the X Window server with a separate media server, or *integrating* the required services into the X Window System. The arguments against integration can roughly be grouped as follows:

- *design purity issues*: The extensions of the system required for multimedia introduce completely new concepts (time, audio) that are quite alien to the original purpose of the system (graphics). However if the term "Window System" is understood to encompass all forms of interaction between applications and user in a modern desktop system, the omission of "audio" appears rather arbitrary and it can be argued that the original design is incomplete insofar.

- *code complexity issues*: Since the X server mediates all interactions between user and applications it is a very critical piece of infrastructure that must not fail; therefore care must be taken that new functionality does not adversely affect reliability and security, and it must be implemented in a way that does not negatively affect software maintenance.

- *implementation issues*: The existing reference X server implementation is not very well-suited for real-time processing at all; since multimedia in general and audio in particular has inherent real-time constraints, this makes it a less than ideal basis for implementation.

The arguments that are strongly in favor of integration are:

- *Protocol closure.* Since intra-stream synchronization requires a common timebase for multiple distinct media (most notably video as well), the protocol(s) must be able to refer to this timebase. This is achieved most easily by making the timebase a "first-class object" of the X Window System.

- *Universality.* The services provided by the extensions to the X Window System presented further down are useful outside the scope of multimedia processing. For example, the ability to submit "post-dated" and precisely timed drawing operations (see section 4.1.1) can also by used for simple animation of graphical user interface elements.

- *Optimization potential.* Several operations required for processing of media can benefit from delegation to specialized hardware. The X server already interfaces with the graphics hardware in sophisticated ways, so it is useful to reuse this infrastructure.

## 4.1. MEDIA PROCESSING EXTENSIONS

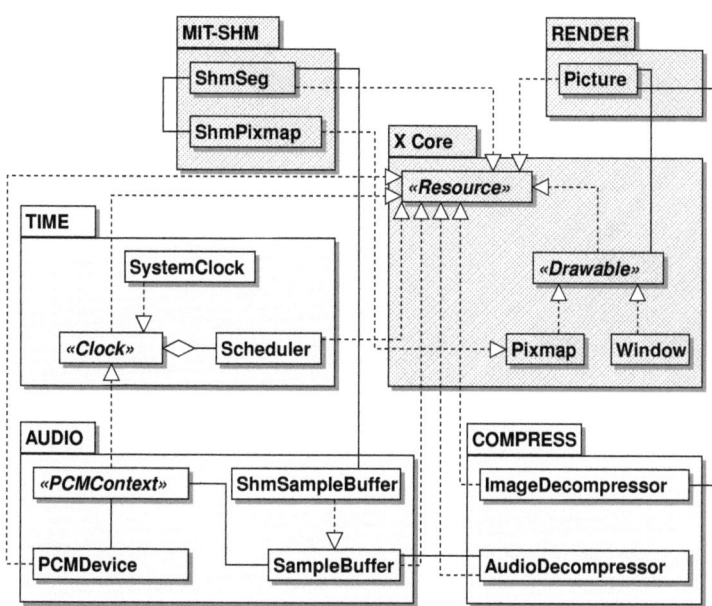

Figure 4.1: Relationship of TIME, AUDIO and COMPRESS extensions to core X services and standardized extensions.

Therefore, the following extensions which complement existing X functionality have been implemented:

- Timing services (TIME extension): add the concept of "time" to the X Window System, introduce clocks into the X server and allow X client applications to issue commands to the X server that are to be executed by the server at a client-defined point in time.

- Audio services (AUDIO extension): allow to process audio in the X Window System in a fashion analogous to the existing image processing infrastructure; cooperate with the timing services to allow audio/video synchronization.

- Compressed media services (COMPRESS extension): introduce the concept of compressed media representation, allow transmission of media data in compressed form, provide an interface for algorithms that deal with compressed media, and allow client-side control over data and algorithms.

Each of these extensions provides functionality that is orthogonal to the others and is useful on its own. Although the extensions have some implementation interdependencies, they are relatively weak. The following three sections describe these extensions in detail. The resource types introduced by the new extensions and their relationship among each other as well as the existing X infrastructure is depicted in figure 4.1.

### 4.1.1 Timing and synchronization services

The basic idea to achieve precise presentation timing is to have the client submit operations to the X server that are however not executed immediately but are deferred to a later point in time. Both the operations themselves as well as the point in time is completely controlled by the client. Timing and synchronization of (multi)media presentation is thus split into two parts:

- The client has to determine the appropriate points in time for all operations to be executed; inter- and intra-stream synchronization requirements must be specified by submitting operations with appropriate timing information.

- The server will just perform the operations the client has requested, at the time specified by the client; specifically the server does not have any explicit knowledge that an operation is part of an ongoing (multi)media presentation, nor does it have explicit knowledge of multiple synchronized media.

This basic idea is the same as set out in [3] and all arguments apply without modification; they will be briefly summarized here for the convenience of the reader.

The timing requirements for (multi)media applications are far less strict than e.g. process control systems. Especially failure to meet a deadline does not result in catastrophic failure of the system but instead gradually degrades the user's experience of the system. Therefore, it is acceptable for multimedia applications to treat the timing requirements as "soft" – i.e. it may "occasionally" fail to meet a deadline.[3] This leads to the following service guarantee for scheduled operations:

(1) Client applications may submit operations to be executed within a client-specified *validity interval*. The server *will* begin execution of the operations not before the start of the validity interval, and not after the end of the validity interval. It *should* execute the operations as close as possible to the beginning of the interval.

While from a conceptual perspective it is desirable to guarantee that execution of operations is finished (instead of begun) before the end of the interval, this strong guarantee is in fact a) unnecessary and b) difficult to implement: a)

---

[3]In other words the system need not "unconditionally guarantee" real-time response; a "statistical" quality of service guarantee is sufficient.

## 4.1. MEDIA PROCESSING EXTENSIONS

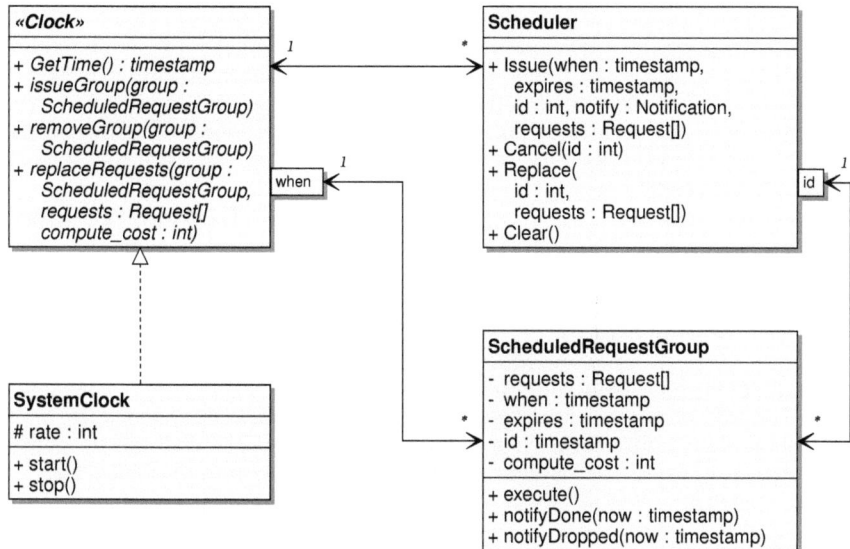

Figure 4.2: Timing and Synchronization services

because in practice execution time is significantly shorter than the validity interval itself, and because b) would require precise knowledge of the duration of the operations.

The operations submitted as a group for a specific validity interval are intended to be used to generate the visible (or audible) output for a short period of time. In this case there is no point in executing the operations only "partially":

(2) Groups of operations submitted by the client are executed with *transactional semantics* – this means that either *all* of the operations are performed, or *none* is. Furthermore, the group of operations is performed atomically with respect to all other operations submitted to the same scheduler.

If multiple groups of operations have overlapping validity intervals, their relative execution order is naturally undefined. The strategy in most real-time systems is to execute pending operations "shortest deadline first"[4]. This conflicts however with the intended semantics of the timing specifications as outlined in (1), so it seems more natural to process groups of operations by their respective begin of the validity interval instead. Experiments so far have failed to establish a clear advantage of either strategy as failure to meet a deadline turned out to be rare occurence, therefore this choice is somewhat arbitrary.

---

[4]The rationale is, of course: if there is *any* schedule at all that guarantees timely execution of all operations, then "shortest deadline first" is guaranteed to provide such a schedule.

The implementation of this conceptual idea requires the introduction of the concept of *time* into the X Window System. This concept is represented within the server as a new class **Clock** which provides the abstract interface that is implemented by all other classes that are able to provide time information. The application-visible interface of this class allows nothing more than request the current "time" from the clock. Other resources that require time information can be "bound" to a clock.

A concrete implementation of this interface is provided in the form of the **RealTimeClock** class; it implements a clock whose notion of time corresponds to wall clock time. It can be instantiated multiple times (so multiple concurrent applications can each have their own private Clock) and each instance can separately be paused and continued.

Support for post-dated operations is provided by the **Scheduler** class. It acts as a "buffer" for all operations that the client wishes to execute at a later point in time, and it takes care of timely execution. It also provides feedback (through X events) to the client whether a specific group of operations was executed, and at what point in time. The relationship of the classes is shown in figure 4.2.

The progression of time represented by a server-side clock can not be influenced by client applications in any way (except for coarse control allowed by the operations of starting and stopping the clock). From the point of view of clients they should be treated as "physical" time sources (in contrast to "logical" time sources which may be sped up or slowed down in relation to physical time) which essentially means that server-side clocks *cannot* be synchronized to a client's clock (or any other clock, for that matter). Additionally, client applications must not assume that different clocks within the server are synchronous.

These limitations are deliberate to simplify the implementation of server-side clock sources[5] and provide deterministic behaviour. This means however that additional work must performed by client applications, see section 4.2.3 for a possible approach.

### 4.1.2 Audio services

Audio has traditionally never been a part of the X Window System. Several networked audio systems have been devised to operate alongside the X server, but despite their complexity most of them have been found lacking even for the simple purposes of presenting synchronized video and audio streams on remote displays. Instead, a comparatively simple and straight-forward extension to the X Window System has been implemented that provides the required functionality. It has been designed in a way to be useful beyond the scope of (multi)media processing – section 5.1 will discuss how other audio APIs can be mapped to the audio extension.

---

[5]It follows the general design principles of the X Window System: "Do not add new functionality unless an implementor cannot complete a real application without it." (Bob Scheifler). See also [55].

## 4.1. MEDIA PROCESSING EXTENSIONS

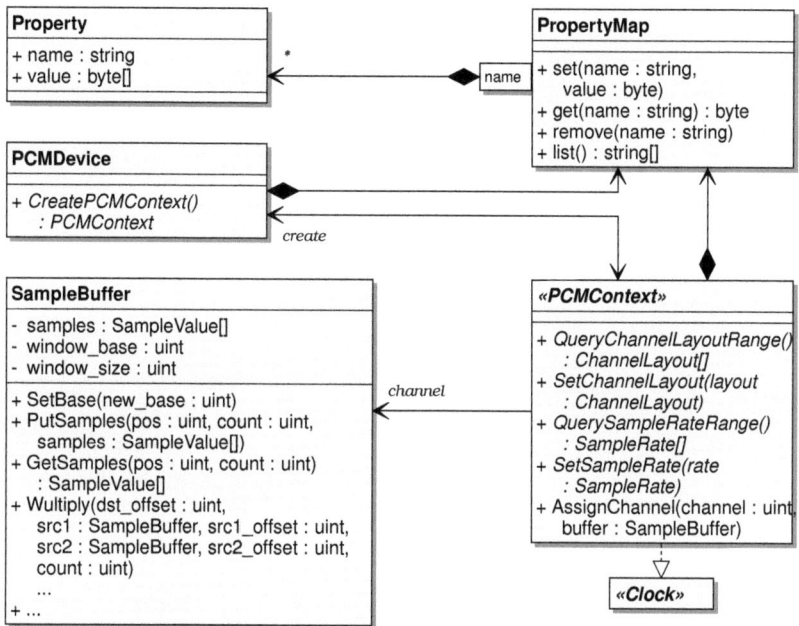

Figure 4.3: Architecture of the Audio extension

The extension presented here follows the general architectural guidelines of the X Window System. It introduces a server-side storage for audio data (a **SampleBuffer**, analogous to a **Pixmap**), compositing operations for server-side audio data (analogous to the core X drawing operations, or the RENDER operations) and server-side objects that specify the interpretation of audio data (a **PCMContext**, analogous to a **Visual** though **PCMContext**s are also used for capture).

The **SampleBuffer**s act as pure "data storage" for samples and lack information (such as sample rate) that would be required to interpret the data as "audio". This information is instead held only in the **PCMContext** objects that require this information to convert sample data to and from analog audio signals. This separation of concepts has a number of consequences: Since the same sample data can be interpreted as audio in different ways (e.g. different sampling rates) the application must ensure that only data with "matching" interpretation is combined. If format conversions are required, the application must *explicitly* perform these conversions; thus the application has complete control over the operations performed on the data, it can e.g. balance computational complexity and quality of the result to its specific needs. This also means that the server-side operations are truly primitives, and they can be implemented very efficiently.

In the same spirit, **SampleBuffer**s contain only data for one channel of audio,

132    CHAPTER 4.  COOPERATION WITH THE X WINDOW SYSTEM

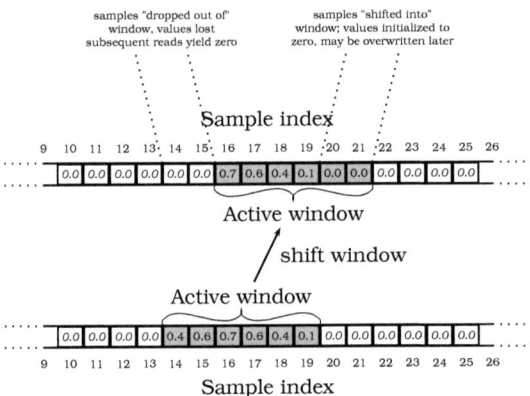

All samples outside the window of interest are implicitly assumed to be zero; sample values that "drop out" of a window are lost, sample values "shifted in" are initialized to zero.

Figure 4.4: Shifting the "active window" of a **SampleBuffer**

so multi-channel audio must be decomposed into multiple buffers. Operations that must be performed on multiple channels of audio must thus be applied to the data of each channel separately, but the flexibility gained allows applications to perform operations such as sophisticated cross-channel mixing.

Figure 4.3 illustrates the relationship of the classes providing audio services.

#### 4.1.2.1  Audio data representation

Audio is represented within the X server as *sampled audio* (in the same sense that the server only deals with *sampled images*). The resource class used to represent samples within the server is a **SampleBuffer**: It stores a set of samples, where each sample consists of an (index, value) pair. A **SampleBuffer** can only represent samples with a contiguous range of indices, but the index need not necessarily start at zero. The sample index is representable as an integer and uniquely identifies each sample, the internal representation of the sample values is left unspecified[6].

Conceptually, a **SampleBuffer** represents a small window of fixed size out of an infinite array of samples. The window can be shifted back and forth so that the range of represented sample indices changes. Only values that lie within the window currently represented by the **SampleBuffer** have meaningful values – any access outside this window will yield 0 (if the sample values are read) or simply

---
    [6]The current implementation uses floating-point values. For computer architectures lacking a FPU it might be preferable to use a fixed-point representation and the concept explicitly allows for alternate implementations.

## 4.1. MEDIA PROCESSING EXTENSIONS

discard the value (if the sample value is written). Shifting the window will also modify the information about the samples: the values of all samples that "slip out" of the window will be lost (and replaced by an implicit zero), and the values of all samples that "slip in" to the window will be initialized with zero (though they can be modified to a different value later). Within the window the samples may be accessed randomly (see figure 4.4).

While this concept of a "sliding window" of samples may appear unusual at first it has the important property that the index of any individual sample is constant and unique. Sample buffers will be used to convey data between a producer and a consumer, and this concept simplifies concurrent access to the data as producer and consumer need only be "weakly" synchronized: This means that producer and consumer can operate largely asynchronous and oblivious of each other; it must only be ensured that a) the producer does not fall behind the consumer and b) both operate within the window of a sample buffer. The chosen data structure for **SampleBuffer**s is conceptually advantageous to other data structures that could be used to implement p/c models:

- A simple array of sample values where the consumer "drains" data from the front (fixed index 0) and the producer "appends" data to the end: random access to the data is possible by both producer and consumer, but access to the data must be strictly serialized; moreover detecting overflow/underflow requires additional bookkeeping.

- A FIFO buffer decouples producer and consumer (although handling of overflow/underflow still needs some work), but severely restricts the access pattern to the data; if multiple operations must be performed on a group of multiple samples, auxiliary buffers are required to represent the intermediate values.

For the common case of client and server residing on the same physical machine, the AUDIO extension also allows placement of sample buffers in shared memory segments (reusing the mechanism already available for pixmaps). For this use case the sliding window concept realized for sample buffers provides another advantage: It can be realized as a lock-free data structure (see appendix B.2).

In practice, a **SampleBuffer** implementation will reuse sample index values as well simply because representation and processing of arbitrarily sized integers is too tedious to implement. But in contrast to a ring-buffer the values are not taken modulo the size of the ring buffer, but instead modulo the number of representable integers which will typically be $2^{32}$. Assuming a sampling rate of 96kHz, such an implementation would thus have to reuse index values after 12 hours, while the before/after relationship of indices becomes ambiguous due to signed overflow for any two sample indices more than 6 hours apart. This "limitation" is assumed to be completely irrelevant for all practical purposes and therefore deemed acceptable[7].

#### 4.1.2.2 Sample data operations

The server can perform several elementary operations on sample data stored in server-side **SampleBuffer**s. Neither do the available operations provide a complete synthesis operator set, nor is this the intended goal. Instead, they are designed to provide sufficient functionality to express common audio *compositing* operations that are performed at the last step before the data is converted into an analog signal; this includes

- up- or down-mixing of multiple audio sources or channels with source or channel-specific weights
- up- or down-sampling
- panning
- simple filtering

Nevertheless, the operations can be (ab)used for very simple audio synthesis. The available functions can be grouped as follows:

- sample up-/download: Enables a client operation to copy client-generated sample data into the server memory or vice versa
- synthesis of simple functions: linear slope, constant, exponential attack/decay, simple periodic functions
- clipping: limit value range to cut away peaks
- arithmetic operations: multiply sample values and accumulate them into a **SampleBuffer** (or replace the old values)
- filtering: convolute two sets of sample values and accumulate them into a **SampleBuffer** (or replace the old values)

The first three groups of operations require no further explanation. The arithmetic operations operate on equally-sized slices of multiple sample buffers, with arbitrary offsets into the individual buffers. The operations supported are:

- `MultiplyConstant(dst, src, constant)`: multiplies each sample value of `src` with `constant` and stores the resulting value into `dst`

---

[7]If producer and consumer manage to become out of sync by more than 6 hours there are *considerably* worse problems to worry about!

## 4.1. MEDIA PROCESSING EXTENSIONS

- `MultiplyAccumulateConstant(dst, src, constant)`: multiplies each sample value of `src` with `constant` and adds the resulting value to the previous value in `dst`

- `Multiply(dst, src1, src2)`: multiplies each sample value of `src1` with its corresponding sample value from `src2` and stores the resulting value into `dst`

- `MultiplyAccumulate(dst, src1, src2)`: multiplies each sample value of `src1` with its corresponding sample value from `src2` and adds the resulting value to the previous value in `dst`

More complicated operations can be expressed by a combination of arithmetic and synthesis. For a example "fade-in" (multiplication of sample values with a linear or exponential ramp) requires synthesizing the ramp into a temporary buffer, and then multiplying the samples with these precomputed values.

The discrete convolution operator is usually defined as

$$out_k = \sum_j in_{k-j} kern_j$$

If $in$, $out$ and $kern$ are interpreted as "sampled" approximations to conceptually continuous functions, then this definition requires that all three functions are sampled with the same sampling rate. The convolution operator described here generalizes the well-known discrete convolution operator to allow all three functions to be sampled differently; it contains the above definition as a special case if the sampling rate of all three functions is equal.

Assume that the signals are sampled at intervals $t_{out}$, $t_{in}$, $t_{kern}$; let the continuation functions of the discrete input and kernel be defined as

$$f_{in(t)} = \sum_k \delta(t - kt_{in}) in_k$$

$$f_{kern}(t) = \frac{((k+1)t_{kern} - t)kern_k + (t - kt_{kern})kern_{k+1}}{t_{kern}} \text{ for } kt_{kern} \leq t < (k+1)t_{kern}$$

In other words, $kern$ is interpolated linearly between sample points, while $in$ is replaced with Dirac impulses at the sample points, scaled to the sample value (see figure 4.5 for a graphical illustration). The output function $f_{out}$ can formally be defined as the convolution of $f_{in}$ and $f_{kern}$:

$$f_{out} = f_{in} * f_{kern}$$

The continuous function $f_{out}$ can then be sampled at intervals $t_{out}$ to produce the discrete output signal. This generalized discrete convolution operator has the following properties:

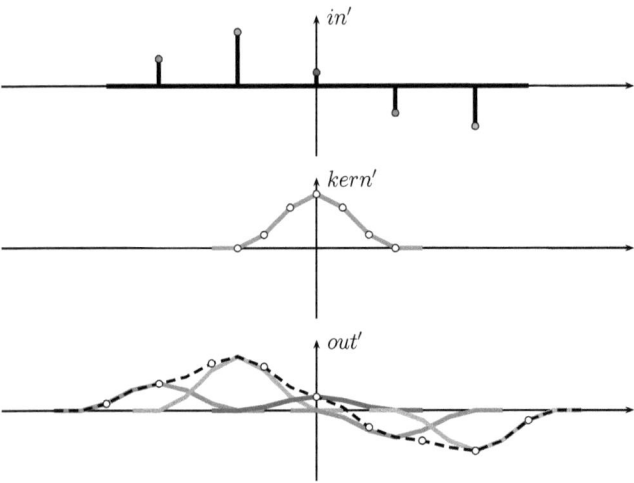

The discrete input signal $in$ is extended to a function $in'$ defined on the continuous time domain that equals scaled Dirac $\delta$ pulses at sample points and zero everywhere else (upper graph). The convolution kernel $kern$ is extended into a function $kern'$ defined on the continuous time domain through linear interpolation between sample points (middle graph). The signal $out'$ is then defined as the convolution of $in'$ and $kern'$ (lower graph); it is a function defined on the continuous time domain, and it may be sampled at desired intervals to produce the discrete signal $out$.

Figure 4.5: Definiton of the generalized discrete convolution operator

- it is identical to the simple discrete convolution operator if $t_{in} = t_{out} = t_{kern}$
- it is generally *not* commutative
- if $in$ and $kern$ are compact[8] with $n_{in}$ and $n_{kern}$ non-zero samples, respectively, then one output sample can be evaluated with at most $6 + 6\max(n_{kern}t_{kern}/t_{in}, n_{in}t_{in}/t_{kern})$ arithmetic operations (the simple discrete convolution operator requires at most $1 + \max(n_{kern}t_{kern}/t_{in}, n_{in}t_{in}/t_{kern})$ operations)

Four variations of this convolution operation are provided, they all operate on slices of three **SampleBuffer**s. Since the **SampleBuffer**s do not have an associated sampling rate or interval, these are provided as parameters to the convolution operation:

---

[8]i.e. non-zero only in a finite interval

## 4.1. MEDIA PROCESSING EXTENSIONS

- `Convolute(out, in, kern)`: convolute `in` and `kern` and place sample values in `out`

- `ConvoluteAccumulate(out, in, kern)`: convolute `in` and `kern` and add sample values to `out`

- `ConvoluteSymmetric(out, in, kern)`: convolute `in` and `kern` and place sample values in `out`; `kern` is treated as symmetric with respect to zero (i.e. $kern_n = kern_{-n}$)

- `ConvoluteAccumulateSymmetric(out, in, kern)`: convolute `in` and `kern` and add sample values to `out`; `kern` is treated as symmetric with respect to zero (i.e. $kern_n = kern_{-n}$)

The "symmetric" variants allow to represent symmetric convolution kernels with half the number of sample values and allow some arithmetic optimizations. Note that the convolution operator is sufficiently generic to implement bandlimited resampling (see 1.3.2.2, 26).

### 4.1.2.3 Audio playback and recording

While **SampleBuffers** act as pure containers of sample data, they provide no means of playing back their contents or capturing from analog sources. This is facilitated through separate **PCMContext** objects instead.

**PCMContexts** reference one or more **SampleBuffers** that are to be used as data storage for either capture or playback, and also include an index per buffer to be used as the "current" playback or capture location. In addition to the data store, they contain all parameters that are required to interpret the underlying sample data as "audio", including sample rate[9], assignment of buffers to channels and intended channel semantics.

Additionally, applications may specify "latency" parameters for each **PCMContext** – for playback this puts an upper limit on the "fetch"-latency that can be tolerated by the application[10], for capture this limits the time between capturing a sample and the point in time that the application may read it. Latency must be negotiated between application and X server, who may impose some technical restrictions.

Once activated, the **PCMContexts** will start reading or writing samples with consecutive indices from or to the respective **SampleBuffers**. However, they will not modify the position of the **SampleBuffers**' windows. This means that they can possibly attempt to access data outside the current window of a **SampleBuffer**. This will result in the behaviour explained in section 4.1.2.1 – output channels

---

[9]Usually, all channels will have the same sampling rate, but for LFE channels lower sampling rates are customary and explicitly permitted by this model.

[10] This latency denotes the maximum time that a sample may be fetched from a data store before it is actually played back. This also denotes the minimum that an application must supply samples ahead of time.

will go mute, and the data from input channels will be lost. It is therefore the applications' responsibility to shift the window of all **SampleBuffer**s in a timely fashion.

Applications therefore need to synchronize with playback and recording to some extent. To facilitate this, **PCMContext**s provide the **Clock** interface as explained in section 4.1.1. Applications can use this interface to query the current playback or recording position or schedule operations accordingly.

#### 4.1.2.4 Audio devices and compositing

**PCMContext**s are instantiated by applications from server-side **PCMDevice**s. These may represent physical PCM capture or playback devices available to the application. Like the screen, physical audio devices are limited resources to which access by multiple parties must be multiplexed. Applications rarely require exclusive access to the full screen, instead the screen is virtualized as multiple "windows" (and a utility application helps the user to manage the windows); likewise applications rarely require exclusive access to the audio device, and a mechanism to multiplex multiple audio is required as well.

Audio recording through a single digitization device is generally not multiplexable since filtering out the "relevant" audio information for a particular application requires significant amount of signal processing and filtering. Therefore, the only form of "multiplex" supported is to replicate the recorded sample data for every application and let the applications filter the parts they are interested in themselves. It may appear problematic that multiple – possibly unrelated – processes are allowed access to the same data; however, since ability to record audio data enables an application to "eavesdrop" on the user, access control restrictions for audio recording capability must be in place anyways. Therefore, the problem of multiplexing access to the recording capability of an audio device is generally best managed within a comprehensive security framework which is out of scope for the present work. The rest of this section will only deal with audio playback.

The model chosen for audio follows the architecture of the RENDER (described by Keith Packard in [45]) and COMPOSITE: In this model, the server allows a "special" client application to "redirect" other applications' drawing operations into off-screen areas; it is then responsible for compositing the on-screen view of all application windows. The server further supports this model with a notification mechanism about window content changes.

For audio this means a **PCMContext** may either be bound to a "physical" PCM device (representing a true hardware device), or a "virtual" PCM device created by another X client responsible for audio compositing. If another application attempts to start playback through such a virtual **PCMContext**, the X server will notify the audio compositing X client which then takes over responsibility for processing of the audio data. It should use the audio compositing operations from section 4.1.2.2 to mix the data from all clients that wish to output audio

## 4.1. MEDIA PROCESSING EXTENSIONS

into its own **SampleBuffer**(s). See section 5.2 for a discussion of this compositing model.

### 4.1.3 Compressed media services

Timing, synchronization and audio services extend the X server's functionality by introducing completely new concepts; they allow to efficiently *process* media data that is already inside the X server process space, but they do not solve the problem of data *transfer*. As outlined in section 4.1 the amount of raw data required to represent a sequence of images and an audio stream may exceed the capacity of the communication channel. It is therefore desirable to transmit the data in compressed form.

Two new resources are introduced to support compressed media within the X server: **ImageDecompressor** and **AudioDecompressor** (cf. figure 4.6). They receive, buffer and decompress media data. The concept of an "image sequence decompressor" was already introduced in [3].

**ImageDecompressor** and **AudioDecompressor** objects are upon instantiation connected to a specific *compression format*. This format essentially identifies the "algorithm" used to convert compressed data into its uncompressed form (and vice versa). "Minor" variations to the algorithm (such as differing parameter sets) can be controlled via *properties*.

- they can manage compressed data stored in the resources (e.g. transmit data)

- they can request conversion of data into uncompressed form

The server side decompressors do not act on unstructured streams of bytes, instead the client has to scan the bytestream representing a video or audio track and decompose it into individual frames (corresponding to one self-contained decompression unit, typically a single image ). These frames are then transmitted to the server, and the client will generate a "handle" to the frame; this handle may be used to refer to the data in later operations, most notably decompression. This means that *transmission* to the X server and *processing* of compressed data within the X server can be performed asynchronously[11], in fact both operations can be performed by different X clients[12]. The rationale for this design has also been covered in [3], so the consequences will only be briefly summarized here:

- Individual frames may be "skipped" by simply not requesting their decompression; no extra server-side processing (scanning of the bytestream) is

---

[11]to some degree: Obviously data must be submitted before it can be processed. To protect the system from "runaway" clients that submit data which is never processed, there is also a limit on how much data can be stored – therefore the data consumer must not lag behind the producer "too much" and both need to find a way to synchronize.

[12]The client submitting the data will however need to find a way to make the handles known to the client wishing to use the data.

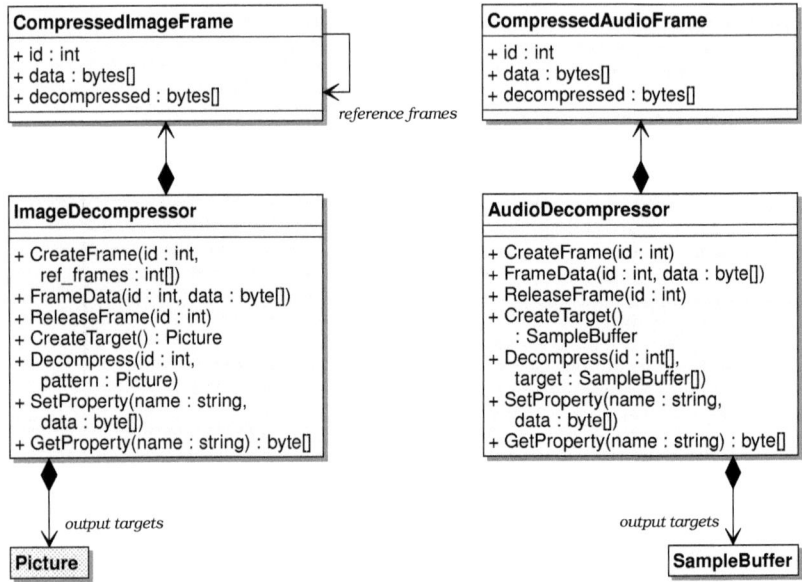

*Classes implementing compressed media services. **AudioDecompressor**s and **ImageDecompressor**s are directly accessible as X resources. They have one or more associated output targets which make the decompressed data available as read-only sources. Data is stored in frame objects that do not have an XID assigned, so they are accessible only through the decompressor object they belong to.*

Figure 4.6: Classes defined by the COMPRESS extension

necessary in this case. Decompression of frames may be requested "out-of-order".

- The system can deal with "missing frames" gracefully and without synchronization problems.

- Decoding dependencies for every single frame must be made explicit; since decompression of frames may be requested out-of-order (with respect to the ordering of frames within a bytestream), the decompressor cannot infer the dependencies implicitly.

Decompressed images and audio samples are available as "source" images for image compositing operations, or "source" sample buffers for audio compositing operations; from the decompressors' point of view these are considered "output targets": every **ImageDecompressor** object has one or more attached **Picture** objects (correspondingly, **AudioDecompressor** objects have one or more attached

**SampleBuffer** objects). These objects are usable as read-only resources for image or audio compositing operations, allowing clients to "blit" image data into other **Picture**s or composite the audio further.

The client may "decompress" audio and image frames through the **decompress** method. The client is expected to specify the output target handles that are to be identified with the decompressed audio or image data. Despite its name, the method does not necessarily immediately *decompress* the data – this step may be delayed until the selected handles are actually used in compositing operations.

## 4.2  Media presentation in the X Window System

The mechanisms outlined in section 4.1 provide the basic primitives that allow media presentation in the X Window System in a network-efficient manner. This section will show what operations a correct media presentation application (using these primitives) will have to perform. We will for the purposes of this section assume that the media consists of one video stream and one audio stream, both given in compressed representation.

The application has to perform the following basic tasks:

1. Acquire resources and initialize required server-side objects

2. Transmit some media data and schedule presentation of data

3. Start presentation as soon as "enough" data has been accumulated in server-side buffers

4. Periodically transmit more media data and schedule presentation of data

5. Compensate for clock drift

This sequence of steps must be performed for each type of media, properly interleaved. The steps will now be explained in more detail, but for the sake of readability the discussion will only deal with one type of media at a time; it is left to the imagination of the reader to properly interleave the conceptually parallel steps.

### 4.2.1  Video presentation

Steps 1 through 3 are illustrated in figure 4.7. It is assumed that a window where the video images should be shown has been created already, and a **Picture** has been associated with this window ❶. The application creates a **ImageDecompressor** object to receive and store compressed image data and manage decompression, specifying the compression format as well as dimensions of the

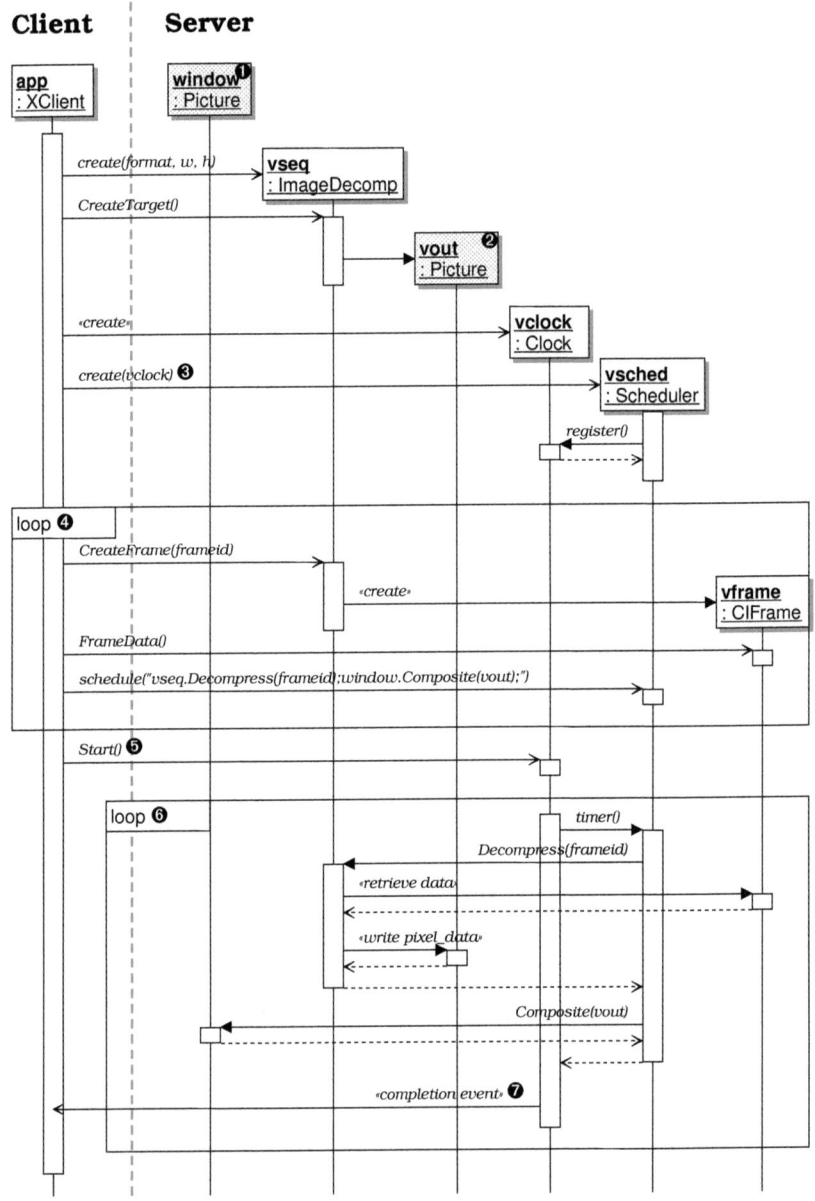

Figure 4.7: Video presentation using the COMPRESS and TIME extensions

pictures within the sequence. Some formats allow or require additional parameters influencing (de)compression to be specified (such as non-standard quantizer matrices for MPEG 1/2 video streams). The application needs to have this information extracted from the video source at this point in time. Furthermore, the application creates a **Picture** handle that will later represent the currently processed image ❷. This handle can then be used to specify a compressed image as a source **Picture** for drawing operations. Finally, a **Clock** and a **Scheduler** are created to provide playback timing, and the **Scheduler** is bound to the **Clock** ❸. This concludes the preparatory step of creating required resources.

Before playback starts, the application submits both compressed data to the image sequence object and timed commands to decompress and display individual images to the scheduler ❹. Submitting image data consists of informing the X server about decoding dependencies (through **createFrame**) and transmission of actual image data (**FrameData**). It is the application's responsibility to extract the required information from the video source and decompose it into individual images.

Finally the application can start the presentation by starting the server-side **Clock** ❺. The server will then autonomously execute the scheduled commands on behalf of the client ❻ which will cause the images to be decompressed and displayed one by one. The server will generate reports of command execution ❼ so that the client is informed about the progress of media presentation. The client can detect drift between its own and the servers clock through the timestamps contained in each report – this will be used for synchronization described in section 4.2.3 below.

Periodically the client needs to supply more compressed image data and schedule new commands for display. Conceptually, the client will execute the same commands as when submitting initial data before starting playback ❹, but the loop has to be synchronized with presentation progress to avoid overflowing or underflowing the server-side buffers.

### 4.2.2 Audio presentation

Transmission of compressed audio frames and decompression scheduling is performed completely analogous to video presentation so only the differences to the process outlined in figure 4.7 will be noted here.

Instead of a picture, the application has to create a **PCMContext** and bind one sample buffer to the context for each audio channel as shown in figure 4.8. The sample buffers will be used in the same way as the window resource in section 4.2.1. The application does not need to create a separate clock since the playback context already provides the **Clock** interface, consequently the **Scheduler** is bound directly to the playback context. Compressed audio data is treated in the same way as compressed video data except for the fact that multiple simultaneous channels must be processed. This means that multiple output sample buffers are created for a single **AudioDecompressor** object (one for each channel),

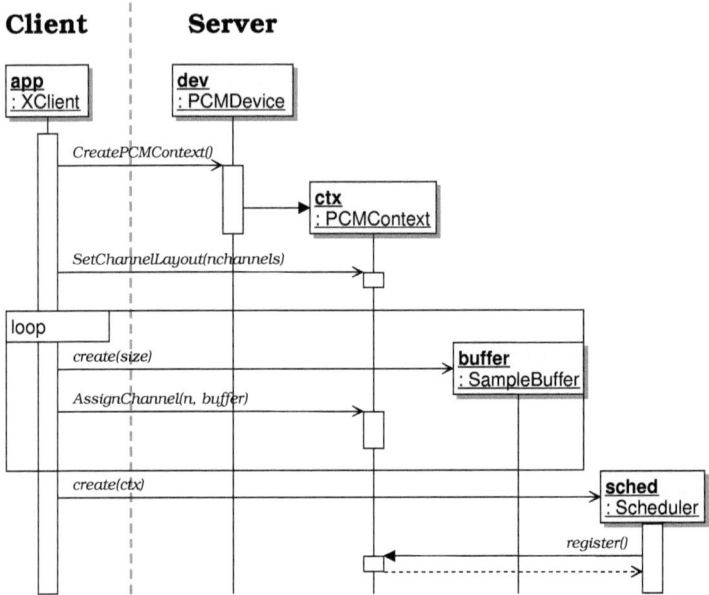

Figure 4.8: Preparation for audio playback

and multiple audio compositing operations must be scheduled at each time step to provide data for the individual channels (figure 4.7, ❷ and ❹).

### 4.2.3 Synchronization

Different clocks may be used for audio and video and they can (and in practice *will*) drift apart so that the presentation will lose synchronicity. In addition to the two clocks just mentioned, there generally is (at least) one more clock involved: the clock used by the client application to determine when to send more audio or video data to the server.

Since the server-side clocks cannot be influenced, the client has to modify the *scheduling* of operations sent to the server. But even though the progression of time within the server-side clock can not be *influenced* by the client, it can be *monitored* by the client. Since the delivery of reports is subject to network-induced delay and jitter, they have to be evaluated statistically to determine systematic clock drift. Additionally, the client can also use this information to determine round-trip delay and create a synchronized "copy" of every server-side clock (see for example the **TickSource** concept described in section 3.2.3.2).

With these information available the client may adopt a variety of strategies to synchronize the presentation. It may choose any clock it wishes as the master clock for the presentation. Adapting video playback to this master clock can be achieved by scheduling the display of images faster or slower than the nominal frame rate, or by skipping/repeating images. Adapting audio playback can be achieved by duplicating or omitting individual samples, or more sophisticated resampling techniques. The extensions of the X Window System provided within this chapter do not force users to choose a specific adaptation strategy but allow applications decide on the best tradeoff between quality/complexity that suits their needs best (see for example the **TimeMapper** service described in 3.2.3.3).

## 4.3 Renderer driver architecture

Services for media presentation are provided as components implementing the **Renderer** interface of libmedia (see section 3.4.3 for a description of the context), and a set of specific components use the X Window System as output target. (These do technically not form part of the core library, but are contained in the separate libmedia-x11 library). The **X11Renderer** and associated components can generally operate in two different modes:

- "legacy" mode: must be used in the absence of the TIME, AUDIO and COMPRESS extensions described above. In this mode the implementation falls back to performing all decompression and compositing operations within the client, sending uncompressed image data to the X server for display and timing all operations within the client. Audio is played back using a secondary audio subsystem. This mode suffers from all disadvantages explained at the beginning of section 4.1 and is only useful if client and server execute on the same physical machine.

- "media-extended" mode: is used whenever possible and TIME, AUDIO and COMPRESS extensions are available. In this mode most of the processing is delegated to the display server, data is transmitted in compressed form, and the timing capabilities of the server are used. Additionally, audio is always played back through the X server.

Since the "legacy" mode is only present as a fallback solution (to make the framework useful with legacy X systems lacking the functionality developed in this chapter), and since its functionality is quite self-evident it will not be discussed here any further. Instead, the discussion in the following sections will be limited to the second mode of operation. Section 4.2 has already shown how the extensions from section 4.1 can be used by media presentation applications. This section will now discuss the **Renderer** and **MediaRenderer** components as well as related classes that together provide a bridge between the framework from chapter 3 and media presentation in the X Window System as is outlined in section 4.2.

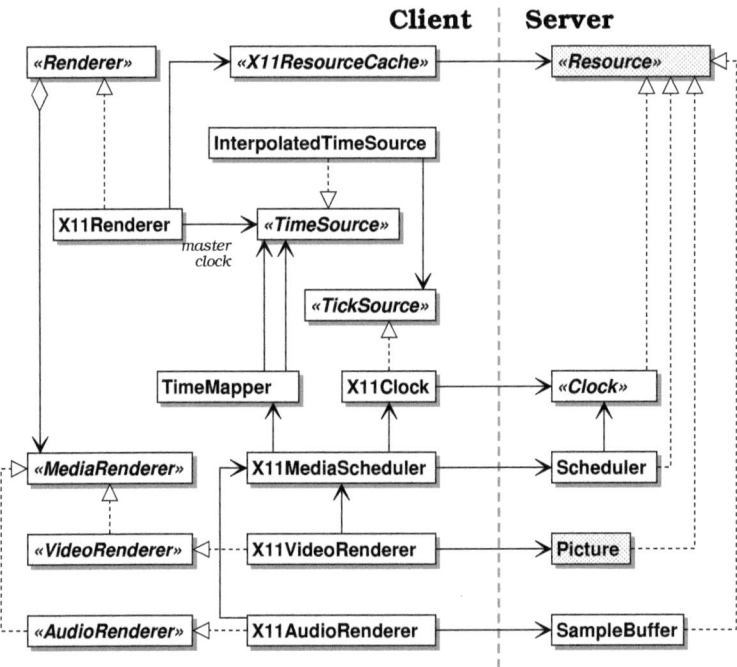

Figure 4.9: General media rendering to an X display, overview class diagram

### 4.3.1 General media rendering and synchronization

The classes involved in general media rendering to an X display are shown in figure 4.9. The central class responsible for coordination and management of all other classes and resources involved is **X11Renderer**. Most importantly it provides a master **TimeSource** which dictates the timing for presentations. All timestamps used to describe media in the context of this renderer will be taken as reference to this time source, consequently all time-dependent media processing operations must be synchronized to it. Any **TimeSource** may serve as master clock – the master time source *may* be identical to one of the other clocks involved, like the X11 audio clock.

Actual image and audio rendering is performed through the **X11AudioRenderer** and **X11VideoRenderer** classes. Both cooperate with the utility **X11MediaScheduler** class which provides the basis for synchronized media presentation. Its main purpose is to provide timed execution of X operations through the mechanism described in section 4.1.1 , but of course the server-side clocks are not synchronized (both to each other and the master presentation

## 4.3. RENDERER DRIVER ARCHITECTURE

clock), so additional work is required to keep the presentation synchronous.

Synchronization is performed according to section 4.2.3 through cooperation of the **X11Clock**, **TimeMapper** and **X11MediaScheduler** classes. The role of **X11Clock** is to communicate with a server-side provider of the X **Clock** interface to receive ticks and provide a client-side synchronized copy of the server-side clock via **InterpolatedTimeSource**. The **TimeMapper** utility provides a mapping between the master presentation clock and the remote clock(s). **X11MediaScheduler** uses this service to map timestamps from the master clock to the remote clocks for execution, so operations can be scheduled for server-side execution accordingly.

After a media fragment has been passed to a renderer, the **X11Renderer** will determine the point in time that the data has to be rendered on the X server (which will usually be the "begin" timestamp contained in the fragment, although the renderer has the option of compensation for known latencies). Having determined the *render* timestamp, it will then determine a *preparation* timestamp: This is the point in time when all data pertaining to the fragment must be transmitted to the server so that it is available in time for rendering.

If the preparation time has not yet arrived, the fragment can be enqueued and the **X11Renderer** will request activation at a later point in time through the master time source. Once the time for preparation has arrived, all data must be submitted. This includes: Compressed or uncompressed audio or image data, scheduled commands that perform required compositing or blits to the screen, and possibly allocation of resources such as required "scratch" pixmaps and sample buffers to hold intermediate data during compositing.

Irregardless of the representation type the data is in, rendering can always follow the basic procedure outlined above: Rendering of sampled media requires placing the media data into server-side resources during the preparation stage; rendering of compressed media is explained in section 4.2 and requires allocation of a decompressor and transmission of the media data; rendering of compressed data must translate the compositing commands to X requests, potentially using temporary resources to store intermediate results.

The **X11Renderer** must keep track of the server-side resources used to represent the media elements. This is done using two principal mechanisms: The first is to statically assign server resources for representation of media elements. The assignment is tracked through the processor-specific data mechanism provided by `libmedia`. This strategy is useful for "long-lived" objects such as decompressors (which are assigned to **CompressedAudioSequence**s and **CompressedImageSequence**s) as well as objects that are "light-weight" in terms of server footprint such as compressed audio/image frames (which are assigned to **CompressedAudioFrame**s and **CompressedImage**s respectively).

The second mechanism, dynamic assignment, is applied to all other media elements: The renderer tries to reuse server-side resources for multiple objects. This is accomplished with the help of the **X11ResourceCache** – its purpose is to keep track of allocated but not statically assigned X server-side objects. This includes scratch pixmaps, pictures and sample buffers used as intermediate storage during complex audio or image compositing operations. While media rendering could in principle be performed without this caching mechanism in place, this would in practice lead to the creation and destruction of many server-side objects in rapid succession. Apart from being inefficient in terms of server processing load it would also quickly exhaust the XID space available to the client.

### 4.3.2 Resource caching

Caching of server-side resources is a common necessity for toolkits interacting with the X server in order to reduce processing load. Immediate-mode drawing toolkits like `cairo` (section 2.2.2.1) can use a relatively simple approach to caching: A resource is either currently in use, or it is not – any resource not currently reserved can be acquired for temporary use, after the operation has finished the resource is immediately returned to the pool for reuse.

Scheduled rendering as performed by the **X11Renderer** also requires resource caching. However, due to the temporal planning-ahead, caching becomes a considerably more intricate matter. Besides the target picture or sample buffer, rendering of a fragment may require several additional resources, including:

a) Resources that contain media pre-computed by the application

b) Resources used for storage of intermediate compositing results

c) Resources used for storage of compositing results that may be referenced by later fragments

These three classes of resources have different reservation life times during which they cannot be used for other purposes:

a) These resources obtain their destined state at the point in time the renderer "prepares" the fragment for playback. Obviously, the state must persist until after the rendering operation has finished.

b) These resources are used only for the duration of the scheduled compositing operation. They may be reused for other purposes at any time before or after this operation. Since scheduled operations submitted to the same scheduler are guaranteed to be strictly ordered, this means that any schedule as a temporary resource for another compositing operation is conflict-free.

c) These resources obtain their destined state during one compositing operation and must persist until after all referencing fragments have been rendered.

## 4.3. RENDERER DRIVER ARCHITECTURE

The **X11ResourceCache** encapsulates the functionality required to keep track of these reservations – which is in fact a surprisingly difficult task[13]:

1. Often *multiple* different candidates can fill in the role of a requested temporary resource – it is for example almost always permitted to use a **Picture** resource that is larger than the one actually needed.

2. The problem is complicated by the fact that the schedule must be dynamic: It is possible that resources of type b) above must be promoted to type c) as the renderer discovers that intermediate data thought to be used only once turns out to be required twice or more.

3. The schedule must be incremental as the renderer has only incomplete information of the future.

4. The cache management must include decisions when to grow or shrink the cache.

### 4.3.3 Handling of media elements

While the procedure outlined in section 4.3.1 covers the temporal aspect of media presentation – scheduling of operations and synchronization – as well as resource management aspects, this section will describe the approach used by the **X11AudioRenderer** and **X11VideoRenderer** to handle actual media elements.

As explained in section 3.4.3, the retained-mode processing model allows some optimizations that can be performed on the media representations before the abstract commands they represent are to be executed – these optimization steps are of course also applied by the X rendering drivers. After any transformations are finished, the renderers must translate the media elements to operations understood by the X server.

As a safe fallback, the renderer driver may always use the **sample** method to obtain a rastered image or PCM audio signal which may then be uploaded into the server[14]. For the various media representation types introduced in section 3.3 it should be quite obvious how they can conceptually be mapped to X operations – for example **SumAudioSignals** of **ProductAudioSignals** can be translated into a single **MultiplyAccumulate** (or even **MultiplyAccumulateConstant**) operation of the constituent signals.

Instead of hard-coding support for the pre-defined media representation types discussed in section 3.3, both **X11AudioRenderer** and **X11VideoRenderer** take a very generic approach: They define two interfaces **X11AudioSignalHandler** and **X11ImageHandler** that provide methods for sending commands to the X server

---

[13]In fact, it is evidently equivalent to "graph coloring" and thus NP-complete, though this theoretical aspect matters little in practice as the number of constraints usually remains well-bounded.

[14]This would quite obviously defeat the purpose of the whole retained-mode processing introduced so far!

to process one media element. Each handler class is implemented for a specific media representation type: For example the handler for **CompressedImage**s is responsible for creating server-side **ImageDecompressor** objects (or using cached instances thereof), transmitting compressed frame data to the server, and for generating commands to decompress and copy the image into its designated destination.

The specific handlers are assigned to the media representation classes using the component and dynamic binding mechanisms outlined in section 3.2.1: Factory objects placed in the namespace **media::x11:mediahandlers** must instantiate the corresponding handler classes. The association is made by giving the factory classes the same public name (in this namespace) as the class representing the media element – thus for a media element of type **SpectacularEffectImage** the renderer would lookup a factory object by the name **media::x11::mediahandlers::SpectacularEffectImage**. The handlers for the built-in media representation types are therefore in no way distinguished from user-defined types (see section 3.3.4) – both are first-class citizens.

# Chapter 5
# System integration

The previous two chapters introduced two software components for media processing and discussed their architecture in isolation[1]. This chapter will discuss how these components integrate into the overall system architecture.

## 5.1 Bindings to audio programming interfaces

The AUDIO extension introduced in section 4.1.2 complements the drawing primitives already provided by the X Window System with audio processing capabilities. The application programming interface provided is in both cases not intended to be device-independent and retargetable, but is specifically tailored to the requirements of networked graphics and audio. While it is possible for application programmers to use these interfaces directly, it is generally undesirable from a software engineering point of view to do so as it couples application logic tightly with the target execution environment.

### 5.1.1 ALSA

In the case of graphics, the drawing functionality (most notably RENDER [45]) of the X Window System is accessible through interfaces as diverse as cairo [10], Qt [16] or GDI/GDI+ (realized through Wine [2]) – which also serves to demonstrate the conceptual completeness of the X drawing model. In the case of audio, ALSA (see section 2.2.1.2) has emerged as the most prevalent low-level interface in use (at the time of this writing). The ALSA plugin that has been created as part of the software implementations for this project enables a broad range of existing audio applications to make use of the X audio extension.

Applications that – in addition to audio processing – also display a graphical user interface already have an established connection to the X server. In principle, it would be desirable to reuse the existing connection for audio, however in

---
[1] Chapter 4 also investigated their interaction.

order for this to work, the application would have to pass the X connection context to ALSA upon initialization of audio operations – which "legacy" operations will not do simply because previously there has never been a reason to do so. Some applications have been modified appropriately, and the modifications have turned out to be simple (though not trivial).

"Unmodified" applications can also be supported by the plugin; this is suitable for applications that do not have an established X connection (and would not need it otherwise), or for "legacy" GUI applications that cannot be modified. In this case, the plugin transparently establishes a (second) connection to the X server that is used for audio exclusively.

The plugin translates ALSA calls to query and configure into functionally equivalent operations on **PCMContext**s and creates server-side **SampleBuffer**s into/from which sample data will be transferred in reaction to the appropriate ALSA functions. Where possible, the plugin transparently attempts to use shared memory instead of transferring the sample data through the X connection (section 4.1.2.1): This allows local desktop applications to achieve latency comparable to direct hardware access (see section 6.2.2).

To recreate the synchronization concept of ALSA it is necessary to provide notifications at the points in time when a sample "period" has passed; audio applications that require low latency typically choose a very low period size accordingly. While the TIME extension provides the necessary mechanism to generate notifications at precise sample times (section 4.1.1), these must be considered unreliable at least in network scenarios as they are subject to jittered delay (as it turned out the reliability is also insufficient in the case of local communication, see section 6.2.2).

Instead, the plugin sets up an application-local timer that is synchronized to the reference clock provided by the **PCMContext**. This timer is then used to "simulate" notification of period transitions.

## 5.2 Desktop audio mixing

Like the physical display, physical audio devices are a resource that must generally be shared between applications that want to interact with a user in a desktop system. The X display model uses the **Window** as an abstraction of "screen area" and provides mechanisms for external entities to manage the "space-multiplex" of windows onto the screen (window manager / compositing manager). Similiarly, the X audio extension discussed in 4.1.2 provides **PCMContext**s as abstractions and mechanisms that allow a separate entity to multiplex access to the physical audio devices.

The model chosen for the AUDIO extension keeps the established separation of mechanisms (arithmetic operations on sample buffers to facilitate mixing) within the X server and policy (decision how audio from different clients are attenuated/muted/mixed) outside of the X server. Figure 5.1 illustrates the data flow

## 5.2. DESKTOP AUDIO MIXING

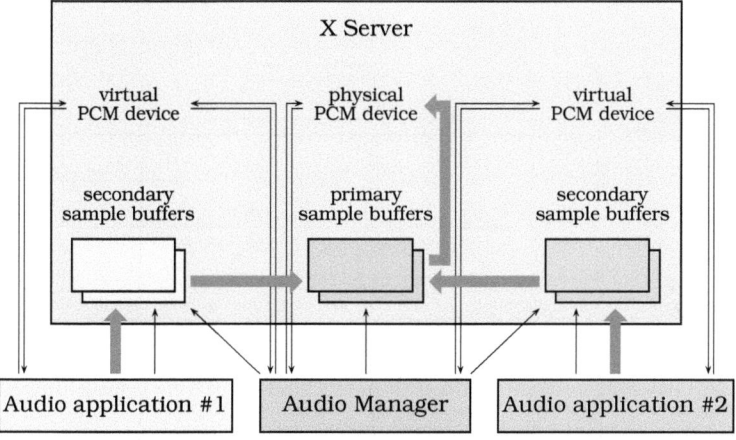

Flow of control (thin black arrows) and sample data (thick red arrows). Audio applications provide sample data in "secondary" buffers (owned by the respective applications) while playback is performed from the "primary" buffers (owned by the audio manager). The audio manager controls how the secondary buffers are mixed into the primary buffer, but mixing operations are performed within the X server so that sample data is never transferred back to the audio manager.

Figure 5.1: Data flow of desktop audio compositing

in this model – one important aspect of this design is that it allows low-latency audio compositing: All sample data is kept locally within the address space of the X server, and all time-critical mixing operations are performed in its context.

Audio compositing follows the model of window compositing (as realized in the COMPOSITE extension) – it is the compositing manager's responsibility to issue X requests that create the desired visual appearance of the screen: The compositing manager may simply blit the application windows into the visible framebuffer unmodified, but may also elect to apply arbitrary transformations on the graphics beforehand. While the window system provides an "automatic" compositing model (where the server simply draws the windows as rectangular areas on the screen), audio compositing does not provide such an automatic fallback path for compositing but *requires* a manager for even the most trivial mixing.

Unlike window compositing, audio compositing must explicitly take the temporal aspect of mixing into consideration. As figure 5.2 illustrates, audio mixing inevitably increases the total playback latency, so the impact has to be kept low in order to meet the demands of applications with low-latency requirements. The audio manager can for this purpose make use of the TIME extension introduced in section 4.1.1; it allows the manager to submit requests that perform the required mixing operations well ahead of time, but have them executed with precise timing.

An audio compositing manager based on the infrastructure presented in

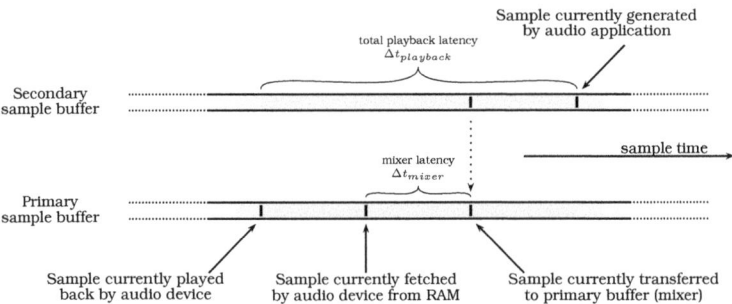

*Audio mixing introduces latencies as samples have to be transferred to the primary playback buffer before the audio device fetches the corresponding samples for playback.*

Figure 5.2: Audio mixing delay

chapter 4 was prepared by Ralf Müller [42], who also developed an extensible concept for realizing different acoustic effects. The audio manager is structured to allow both low latency mixing, but keep coupling of the X server and the manager limited using the timing mechanisms above: The mixing commands are scheduled *speculatively* for a considerable amount of time in the future, assuming that none of the mixing parameters will change over time. When any change occurs (e.g. the volume of an audio stream is to be reduced), the previously scheduled commands are revoked and replaced with different mixing operations.

## 5.3 GUI toolkit cooperation

Applications that wish to provide a graphical user interface will rely on an appropriate toolkit for this purpose. Each toolkit provides numerous services, including a library of graphical control elements ("widgets"). The different toolkits are structured likewise in a number of ways:

(1) The toolkit requires the application to adopt an event-driven programming model; the event dispatching loop is provided by the toolkit itself, applications must register any event sources (e.g. network connections) with the toolkit to receive callbacks.

(2) The control elements ("widgets") are implemented as a hierarchy of classes, derived from a toolkit-specific base class.

## 5.3. GUI TOOLKIT COOPERATION

Using any graphical user interface toolkit together with the media processing toolkit described in chapter 3 and possibly the X rendering driver described in chapter 4 in the same application requires some coordination:

(1) Many of the core or component classes are "active" in the sense that they require execution control in reaction to external events. These are for example **Clock**s, most real-time **Source**s such as those used for capture (see section 3.4.2) or receiving network streams.

(2) The **X11Renderer** driver must share several resources with the toolkit as well as coordinate processing of X events required for communication with the X server extensions.

While (1) could be addressed by deferring the processing to a separate thread of execution, this does not solve issue (2) and moreover forces a specific programming model on the application developer. Instead, the choice has been made to provide an abstract "bridge" to event registration services provided by toolkits (which still leaves the option of using a secondary thread open to the programmer).

Unfortunately, it is not possible to provide these coordination services in a way that is completely independent from the GUI toolkit: The resulting glue code would have link-time dependencies on all thinkable toolkit libraries, and make conflicting event registrations to each of these. The architecture nevertheless provides some assitance which at least minimizes the size of the adaptation layer required for each toolkit. The following sections will give a short overview of the general approach and two examples involving commonly used GUI toolkits.

### 5.3.1 Media framework provisions

The architectural support for toolkit cooperation is split in two parts:

- Generic operating system event service wrappers, provided by the core `libmedia` library.

- X11 Window System event service wrappers, provided by the auxiliary `libmedia-x11` library that also contains the **X11Renderer** components.

The interfaces **timer_service** and **ioready_service** provide client objects with the ability to register for notification at specified points in time, or on I/O readiness detection on a descriptor (representing an operating system object), while **xevent_service** allows notification of X events. Any object requiring activation does not use the operating system or X services directly, but relies on the abstract service interfaces. This also means that these interfaces must be passed to the corresponding objects on instantiation.

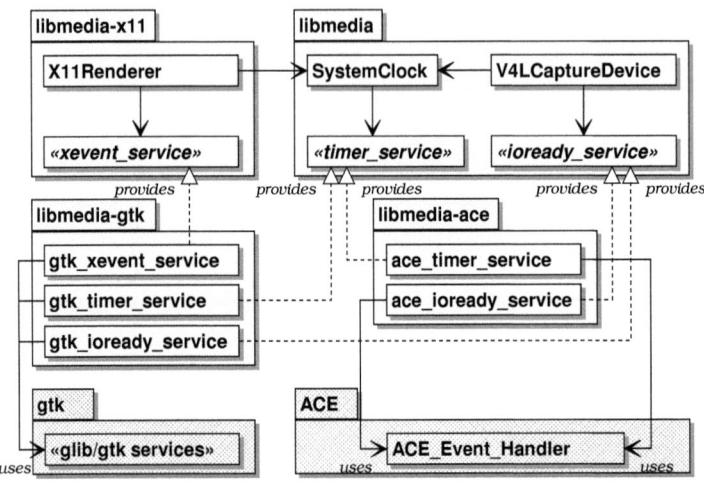

The diagram shows the relationship of the abstract event notification services introduced in `libmedia` and `libmedia-x11`, as well as their relationship to implementations provided in glue libraries (adaptor pattern). Besides the typical use-case of binding to user interface toolkit libraries this also allows binding to other application frameworks such as ACE.

Figure 5.3: Event processing cooperation

The interfaces are implemented as *adaptors* by glue libraries external to the core `libmedia` or the `libmedia-x11` library which in turn delegate the requested operations to services provided by other frameworks (see figure 5.3): This avoids implementation or link-time dependencies on other frameworks. Note that this is strictly speaking not tied to GUI toolkits at all – the library can be bound to other application frameworks such as ACE [60] using the same approach.

Both X GUI toolkit and the media processing framework need to keep track of resources allocated in the X server. No problem ensues as long as each of the frameworks has "exclusive" ownership of the resource, however as soon as a resource is shared suddenly both frameworks are keeping track of it. To avoid inconsistencies, both must therefore be informed that there is another user of the particular resource that is external to the framework itself. For all resources that might be shared `libmedia-x11` provides "tracking containers" to assist in coordinating resource tracking: The containers act as "proxies" that signify usage of a resource by one framework to the respective other. Note that they only assist in *tracking*, but do not provide any indirection (i.e. *wrap* the resource): Any access is direct and without overhead. Appendix A.1 provides more background information on how this "tracker" concept is used throughout the architecture.

## 5.3.2 Gtk+/Qt bridge libraries

Gtk+ provides a rich framework for the development of graphical user interfaces ([37], [23], [48]). While it is conceptually cross-platform and agnostic to the undelying graphics system, it is most commonly used in the context of the X Window System. It has close ties to the cairo graphics library (see section 2.2.2.1) that provides the underlying infrastructure for drawing of graphical user interface elements.

The Qt toolkit fulfills a role that is very similar to Gtk+, and it is the second popular framework in use for realization of graphical user interfaces in the X Window System [5] [8]. While both libraries provide completely different object models and APIs, they are structurally quite similar so that the approach taken for implementation of bridge libraries is largely the same. It should be noted that despite their similar role and structure, there are enough technical differences that the two libraries cannot be used interchangeably.

Both GUI toolkits force the application to relinquish the main control flow to the framework, the programmer must adopt an event-driven application model. Qt and Gtk+ provide large collections of graphical control elements (called *widgets*) organized in a hierarchy of classes. In addition to providing the visual appearance and reactions to user input for an individual element, they must also perform several administrative duties such as interfacing with the layout manager and the event routing system.

A bridge library between libmedia, the libmedia-x11 renderer library and the Gtk+ framework was implemented by J. Pfeiffer [49]. It provides mappings of the event notification services as discussed in the preceding section, as well as a convenient widget that can perform the role of a "viewing area" for video content. A similar bridge library towards the Qt toolkit was also created independently by the author of this work.

Mapping of the generic operating system notifications turns out to be relatively straight-forward as the GUI libraries provides almost exact equivalents of the required interfaces. For processing of X11 events on the other hand, both generally assume that each event relates to a specific window (which for events corresponding to the TIME and AUDIO extensions is not the case) and attempts to route the event to the widget(s) contained in this window. Both toolkits provide a mechanism to install a global event filter[2] that can intercept X11 events before they are routed by the GUI toolkit. The bridge libraries use this facility and create a secondary event routing system to deliver the required notification to the libmedia consumer classes.

The Gtk+ and Qt bridges also provide a "widget" that interfaces with the **X11Renderer** on the one hand and the Gtk+/Qt widget class system on the other hand. Specifically, it allows to hand over X resources created and wrapped by

---

[2]Multiple filters may installed which get a chance to processes an event if none of the previous filters has consumed it yet. Implemented properly, the different global filters will not conflict, but a more fine-grained registration model allowing a consumer to register their interest in specific events classified by "type" and "resource XID" would be preferable to detect potential conflicts.

Gtk+/Qt to the **X11VideoRenderer** class, so that video images can be shown in view ports created and managed by the respective GUI toolkit. In return, the toolkit's geometry manager is supplied with feedback as to appropriate proportions suitable for the video view port, so that it can arrange the overall layout of the window accordingly. Lastly, the implementation of the graphical control elements also provides several high-level media control mechanisms: These applications to initiate, pause and resume autonomous playback of stored media content using a single function call – all required event registration and transport of media **Fragment**s from the media **Source**s is handled transparently by this function. Essentially, this allows to implement a media player application in just three calls: The first to instantiate the controller widget, the second to place it within the rest of the user interface, and the third to initiate playback.

## 5.4 Cooperation with other media frameworks

The media processing architecture presented in chapter 3 has some functional overlap with other media processing frameworks and singular libraries already present for Linux operating environment. While none of the other libraries provides such comprehensive service like the compositing facilities offered by `libmedia` for audio and images, functional duplication exists at least in the handling of compressed media data. Since the implementation of compressors and decompressors is the most arduous (and at the same time conceptually most uninteresting) part of the development, it is desirable to find means of reusing the implementation work done in other projects.

The architecture model chosen in chapter 3 leads to a separation of compressed media processing into two parts. For the processing of initially compressed data, fhe first half is performed by the **MediaSource** instances (or their helper classes): They scan the compressed representation to extract any required meta-data such as temporal duration of a video image or a sequence of audio samples, image dimensions, decoding dependencies and suitable sample formats that can losslessly represent the media. The second half is fulfilled by decompressor objects (see section 3.3.3.2): They perform the actual job of transforming the data into a different representation, and they rely on the meta-data extracted in the first step. For compression and storage there is a similar split into the sequence managers and compressor instances (see section 3.3.3.3).

This concept of split processing is however quite alien to most of the existing compressor and decompressor implementations which generally expect to process both data and what `libmedia` would consider meta-data in a single pass. While there are a few cases of libraries which export an interface that is sufficiently fine-grained to support this two-part processing model (e.g. `libjpeg` [64]), the majority of implementations is unfortunately not very amenable to this approach.

The implementation of the meta-data extraction part of media processing is in general a rather trivial exercise, so this alone would not be a significant obstacle

## 5.4. COOPERATION WITH OTHER MEDIA FRAMEWORKS

in reusing existing codec libraries. However, most compressors and decompressors expose an interface that corresponds to the "streaming" data paradigm also found in filter graph architectures: The codec instances hold *implicit* state to resolve decoding dependencies, while the codec interface of `libmedia` requires that compressor and decompressor instances operate with the *explicitly* provided state information given by the caller.

This semantic mismatch between the codec interfaces makes wrapping existing compressor or decompressor implementations infeasible. Consider for example a decompressor for MPEG-1 video (see section 1.4.3.2, and in particular figure 1.15 on page 47). The stateless **ImageDecompressor** would then be used in the following way:

1. The data for the first image (an I frame) is given to the decompressor instance which returns a decompressed representation of the image. Afterwards this image will be displayed.

2. For the second image (a B frame), the **Renderer** driver would detect the decoding dependency to the first and fourth frame (a P frame) from the **CompressedImage** object. It will therefore resolve all unsatisfied decoding dependencies by requesting decompression of the fourth frame (using the already decoded first frame as reference image). The **ImageDecompressor** will return a **PixelImage** containing the decoded image.

3. Afterwards, the renderer will request decompression of the second frame, using the first and fourth as references. Afterwards this image will be displayed as well.

On the other hand, a decompressor that performs reordering of frames decoding to display order through implicit buffering would behave in the following way:

a. The data for the first image (an I frame) is given to the decompressor instance, but *no image would be returned* as output since the decompressor cannot decide whether the image must be delayed since it does not yet know if the subsequent image is a P- or B-frame (consider the ordering of frames 7 and 5!)

b. The fourth frame (a P frame) is given to the decompressor, the decompressor would return the buffered I frame as decoded image and delay the fourth frame.

c. The second frame (a B frame) is given to the decompressor. In this case, the decoded second picture will be returned without any delay (while frame number four is still kept and delayed).

Wrapping the interface of the stateful, implicitly reordering decompressor instance in a stateless `libmedia` **ImageDecompressor** object is therefore not possible – the operation already fails at step 1: The application may not have given any of the subsequent fragments to the renderer – the data for the second and fourth frames are unavailable. Consequently, the required first image remains "stuck" inside the stateful decompressor as it will only the decoded image after being fed a valid P frame as in step b. above.

It should be noted that providing wrapper interfaces the other way around (i.e. to make the codec implementations conforming to the interface required by `libmedia` usable by other frameworks) is rather trivial: It just amounts to "hiding" the more fine-granular interface.

The various codecs provided as part of this implementation were therefore either written from scratch (e.g. MPEG audio), could rely on individual libraries providing a sufficiently fine-granular interface for specific formats (e.g. JPEG and Motion JPEG), or are the result of heavy refactoring of individual codec implementations (e.g. MPEG-1/MPEG-2 video). While the third approach may appear to be an acceptable approach as it is conceivably easier to start with a known-working implementation, the experience gained during the project is that even this task is far from trivial (see section 6.1.1.2).

# Chapter 6

# Assessment

The various parts of the framework presented throughout the preceding chapters have been implemented on a Linux/i386 system over the course of about 5 years at TU Freiberg. During this period, it has been showcased multiple times in part or in full for occasions as various as the Cebit tradeshow 2007, or open source development meetings. It has been used as the basis for three bachelor's theses, several smaller student projects and course work assignments.

The implementation is stable and has reached the level of maturity where it is usable for the development of practical multimedia applications. Many parts of the implementation have already been published for download, the remaining parts are expected to follow by the end of 2008. Considerable time has also already been spent on optimizations where useful, mostly affecting decompressor implementations critical for real-time playback: the implementations are en par in terms of efficiency with other popular codec libraries such as ffmpeg and not worse by more than about 15% than the fastest implementations known to the author.

The implementation is split into a number of different libraries and in total consists of roughly 80000 lines of code. These divide into:

- libmedia: The core media processing library realizing the concepts presented in chapter 3. It consists of roughly 32000 lines of C++ code.

- TIME and AUDIO extensions for the X Window System: Extension module for the X.org reference X Server implementation as well as required client library for the services described in sections 4.1.1 and 4.1.2. Altogether, this consists of roughly 13000 lines of C code.

- COMPRESS extension for the X Window System: Extension module for the X.org reference X Server implementation as well as required client library for the services described in sections 4.1.3. The server extension module also contains a shim layer to make compressors conforming to the interface defined by libmedia accessible. The implementation is about 5000 lines of mixed C/C++ code.

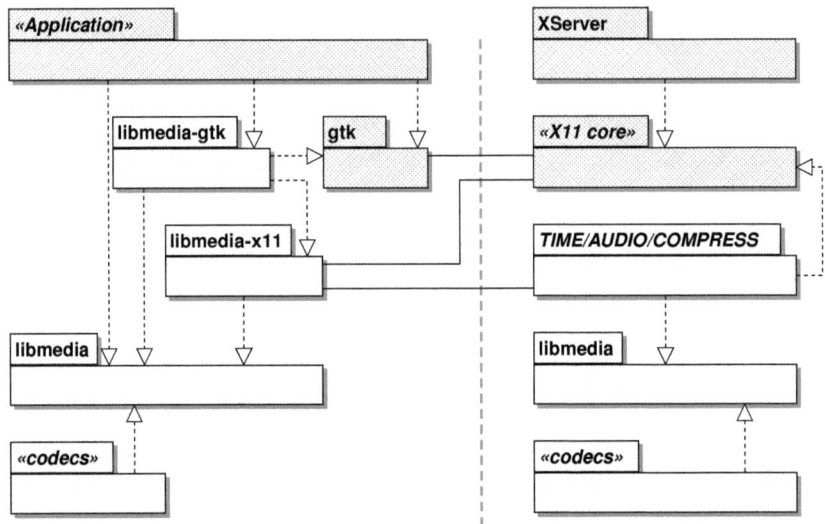

Figure 6.1: Organizational diagram of the different framework parts

- `libmedia-x11`: Renderer driver for the X Window System that can make use of the above extensions (see section 4.3), implemented in about 4000 lines of C++ code.

- `libmedia-gtk` and `libmedia-qt`: Bridge libraries between the X rendering driver above and the respective user interface toolkits (see section 5.3). Each library consists of around 1000 lines of code.

- Various compressor, decompressor, document handler and other components providing functional services for the core library. These add up to about 10000 lines of mixed C/C++ code, but about a quarter of the code was not written from scratch but borrowed and refactored from other projects. (The number does not include the source code of linked-in codec libraries that turned out to be usable without modification).

- Various utilities, such as event dispatching helpers, callback and synchronization management, just-in-time code generation, vectorized multimedia processing kernels etc. totalling about 25000 lines of code.

Figure 6.1 gives an overview of the relationship between these framework parts and other existing system components.

6.1. ARCHITECTURE MODEL AND API ASSESSMENT

This chapter will discuss the experience gained from the implementation. The first section 6.1 will concentrate on the architecture model, the implementation and the resulting API: While the internal model is certainly more complex, the external interface exported by `libmedia` is quite easy to use and noticeably more powerful than that of the frameworks discussed in section 2.1 and allows programmers to achieve more with less code. The next section 6.2 focuses on (quantitative) performance aspects. In particular, it shows that the overhead introduced by the considerable amount of bookkeeping required by the retained-mode model is negligible in comparison to the actual data transformations typically required in multimedia processing applications.

## 6.1 Architecture model and API assessment

The architecture introduces several concepts that are not found in previous architectures:

- The concept of *retained-mode processing* used in this architecture is unusual in the field of time-based media and multimedia processing. Previously, it has already been applied in image processing applications that require an explicit representation of processing steps to be applied for repeatability [63]. It has also been found useful in distributed rendering applications as it lends itself quite naturally to delegated processing [47].

- The distinction between *processing* and *representation* is much more rigorous than in previous architectures. This is manifest in a strict separation of objects that merely represent some media data (e.g. **Document**s, **Image**s, **CompressedAudioSequence**s) but provide no functionality to actually manipulate the data, and objects that modify or otherwise access the data (e.g. **DocumentReader**s, **RowIterator**s, **AudioDecompressor**s). Variants of this design pattern can be found in many other types of software (e.g. the document/view or model/view/controller patterns found in most interactive applications featuring a graphical user interface also strive to make this separation), but at least the discussed existing multimedia frameworks do not apply this principle rigorously.

- The combination of the above two concepts leads to the principle of *lazy evaluation* that is widely known in functional and logic programming, but is also novel to multimedia processing. Here, it is used as an utility to "wrap" the generation of the retained-mode intermediate representation through a more easily understood immediate-mode API. It also introduces an interesting "double indirection" in providing modularity to media processing: New processing operators are implemented in a filter graph architecture by implementing a *single* component that both represents the operation as part of the graph as well as knows how to compute the effect. In contrast, `libmedia` separates these tasks into different components: One that serves

164                                              CHAPTER 6.  ASSESSMENT

as "placeholder" for the operation to be performed in the retained-mode representation of the processing chain, and multiple components that realize these operations in the context of specific renderer drivers (one per renderer driver).

These principles naturally introduce considerable complexity into the architecture, so the natural question is what improvements can be gained over other approaches, and at what costs.

### 6.1.1  API field testing

Since the central goal of this work is to provide a multimedia framework suitable for realization of practical applications, one of the most important criteria for assessing whether the stated goals have been achieved is to create software that uses the interfaces in a meaningful way. The framework is designed with the explicit purpose of extensibility, consequently there are two types of interfaces that must be considered:

- The *external* interfaces are used by applications that wish to perform media processing operations. These are what most application programmers will be exposed to, so convenience is a high priority.

- The *internal* interfaces are used by extension modules that expand the capabilities of the framework, for example by new functional processing components such as codecs. These are not generally used by application developers, they are necessarily more tightly coupled to the overall architecture and are therefore likely to be more complex.

Several example applications and extension modules were created by both the author and other voluntary test subjects[1] during this project, and the results of these experiments will be summarized in this section.

#### 6.1.1.1  External interfaces

To exercise the capabilities of the API and test their suitability for the development of multimedia processing, several sample applications were created by the author. These include:

- "Media player": A simple application that plays back multimedia content stored in a container file, DVD or received as e.g. an http stream. The application interfaces with the X Window System to provide both a graphical user interface for control, as well as the target system to display the content on. This application mainly exercises the **Document** abstraction to access various forms of stored content as well as the **X11Renderer** mechanism for network-transparent playback on a remote X display.

---
[1] a.k.a. "students"

## 6.1. ARCHITECTURE MODEL AND API ASSESSMENT

- "Bulk video processing": A simple command-line tool that can non-interactively read in stored media data, perform several simple processing operations, and write the result back to a container file.

- "Video editing": A simple interactive video editing application that allows the user to open several video clips, cut them, arrange them on a temporal scale and write the result to a container file. Some simple scene transition and other effects are built-in and hard-coded, but the underlying mechanism is conceptually very powerful as it relies on the cairo graphics API – bindings to a scripting interface could vastly enhance the capabilities of this simple application. It provides a graphical user interface through the X Window System and also uses the **X11Renderer** mechanism to allow interactive preview of the edited clip (using the compressed data transmission mechanism as well as the processing capabilities of the X Window System).

The applications themselves are surprisingly simple – except for the video editor they are realized in a few hundred lines of code. The video editor application is slightly more complex due to the GUI interactions, but the media processing core is very compact thanks to the abstract API provided by libmedia: In particular, the **Renderer** concept allows to keep a single code path for both "preview" and "final" compositing of the media elements – while the **X11Renderer** delegates all processing to the remote display, the **DocumentWriter** implementation performs equivalent processing steps locally to store the resulting video clip into a container file.

Field testing of the API by other developers was mainly done by students during course work at TU Freiberg. Overall, the API was well-received and easily understood, in particular the generic **Document** and **Fragment** concepts turned out to be very intuitive. The processing and I/O model provided by the framework allowed the students to quickly jump into programming – for example, the following code sequence is all that is required to extract images out of a video container file[2]:

```
/* open document stored as file in the file system */
ref<Document> doc=openDocument("sample.avi", O_RDONLY);
/* select first video track and open read accessor */
Document::TrackList tracklist(doc->select<VideoMedia>());
ref<DocumentReader> reader=doc->openReader(tracklist);
/* select video reader from opened read accessor */
ref<VideoSource> vreader=reader->select<VideoMedia>();

/* read media fragment and extract image */
ref<VideoFragment> fragment=vreader->getVideoFragment();
ref<Image> image=fragment->getImage();
```

[2] Here and in the following programming examples, ref designates a reference-counted smart pointer. It is functionally equivalent to the **intrusive_ptr** provided by the **boost** library and is a candidate for inclusion in the upcoming C++0x standard.

```
/* coerce conversion into specific representation */
ref<PixelImage> pimage=image->sample(pixelformats::ARGB32_native);
```

The above sample code compares quite favorably to the effort required for the same effect using QuickTime or DirectShow – for example, in QuickTime the equivalent of just the first line of above code would be:

```
/* convert string into representation required by QuickTime
file system utilities */
CFStringRef cfpath=CFStringCreateWithCString(NULL,
   "./sample.avi", CFStringGetSystemEncoding());
Handle dataRef=0;
OSType dataRefType;
/* open data storage designated by file name */
OSErr err=QTNewDataReferenceFromFullPathCFString(cfpath,
   kQTNativeDefaultPathStyle, 0, &dataRef, &dataRefType);
Movie movie;
short resource_id=0;
/* open Movie contained in data storage */
err=NewMovieFromDataRef(&movie, newMovieActive,
   &resource_id, dataRef, dataRefType);
```

while the full example adds up to almost 100 lines of code[3]. While the core processing concepts (lazy evaluation, retained-mode processing) also turned out to be easily understood, the students surprisingly had considerable difficulties with object lifetime tracking – while the automatic reference count tracking facilities provided by the **ref** template were easily understood, the manual tracking required during creation of new media elements was not. Partly, this may be owed to the students' evident unfamiliarity with the concept of reference tracking (which caught the author somewhat by surprise during the trials).

Further comments were received from third party developers during conference meetings and presentations. While the model itself was generally easily understood, it was not unanimously received without skepticism – several of the commenters were concerned about the overhead incurred by the processing model (see section 6.2.1), while others pointed out that it represents a fundamental shift especially with regards to video processing: All of the developers

---

[3]But it must be admitted that part of the brevity of the libmedia code example is owed to the generous use of convenience functions – for example libmedia like QuickTime also provides an indirection layer that decouples the **Document** from the methods used to access the underlying byte storage, so the "full" call to instantiate a document without any convenience wrappers would rather be:
```
ref<ByteSequence> bytes=File::open("sample.avi", O_RDONLY);
ref<Document> doc=openDocument(bytes, O_RDONLY, &mimeTypeClassifier);
```
In the author's opinion, this is much more legible, still. Considerable coding overhead is incurred in QuickTime by the gratuitous unpacking and repacking of data into different structures between calls to different QuickTime API functions.

## 6.1. ARCHITECTURE MODEL AND API ASSESSMENT

were used to the buffering and planning ahead required for audio processing, adoption of a retained-mode processing model would be a rather minor change. The same cannot be said about video as immediate processing of each image just before it is shown or stored is the prevalent model. However, this was not seen so much as a "limitation" of the approach taken by libmedia, but rather an observation that a switch-over would often be architecturally invasive and therefore not taken lightly.

#### 6.1.1.2 Internal interfaces

The internal interface for registration of extension modules is rather straightforward: It requires the developer to

- Create the desired functional component by deriving from one of the many available interfaces. For example, a component that allows embedding of video in the "theora" format into AVI containers, a developer would derive a new class from the **AVIFrameHandler** interface.

- Create a factory class and instantiate a single object as a global variable with proper name so it can be found by its symbol name. In the above example, a developer would declare a global variable by the name **avi::mediahandlers::vids::VP31**, as VP31 is the code used in AVI files to denote this format.

Several extension modules were also created by both the author and students, mostly to provide support for new compressed media representation formats. Generally, the internal interfaces were well-understood and the many auxiliary concepts (such as the **BufferWindow** concept, see appendix A.3) were deemed helpful [6]. In particular, the experience showed that students had little trouble implementing for example the "front end" part of handling compressed media representations (i.e. the scanning and extraction of required meta-data into suitable libmedia data structures).

In multiple student sub-projects it was attempted to convert various existing media codec implementations into decompressor and compressor components for libmedia (cf. section 5.4). Except for two projects these attempts have generally been failures: In almost all cases the students were unable to develop a sufficient understanding of the existing codec implementations – arguably, these are generally complex pieces of code due to the many performance optimizations applied. The successful students reported that understanding the respective media format and code base to work with was far more time-consuming than the library-provided interfaces to be implemented.

Drawing conclusions from both the successful and the failed projects allows to cautiously state that the internal component interfaces are useful and can be understood quite well. However, these projects also illustrate that implementation of any complex media format (either from scratch or through refactoring) is far beyond the capability of average students. This hints at a possible problem with the concepts introduced by the media processing framework that is less of a technical but more of a "human resource" nature: The architecture model prohibits simple reuse of an existing code base, but refactoring existing implementations is a task that on the one hand requires high qualification, on the other hand is conceptually "uninspiring" as there is next to no architectural work.

### 6.1.2 Comparison to QuickTime

Both QuickTime and `libmedia` share an imperative programming model that provides application programmers with a large amount of control: Since there is no implicit "flow" of data, it is ultimately the application's responsibility to specify all individual processing and transformation operations. The implementor controls how media elements are passed through the different functional services provided by the framework.

This similarity means that applications performing comparable tasks are generally also similar in structure, but `libmedia` introduces several concepts that QuickTime is lacking (see the short listing at the beginning of this chapter as well as the introductory discussion in chapter 3). The following sections will therefore provide prominent examples of how the new concepts positively affect the API and to what extent this assists application programmers.

#### 6.1.2.1 Data model and abstractions

The most fundamental difference can be found in the abstractions provided by the respective media frameworks: While QuickTime features a rich set of interfaces that are implemented by various components to make their functional services accessible to multimedia applications, it provides next to no abstractions for the fundamental data elements itself – for example, there simply is no unified abstraction for the concept of an "image" in QuickTime: In the various places where an image is used, an API function may require it to be represented as an "array of pixels" (for rastered images), a device image handle (for images passed to the display subsystem) or an opaque sequence of bytes (e.g. for compressed representations) depending on the context.

Even in cases where objects may superficially be regarded as data abstractions, these usually commingle both representation and processing concepts: For example, a QuickTime **Movie** on the one hand represents a container of multimedia data. On the other hand, it also serves as a context for playback as it has an associated graphics context (to which the video contained in the **Movie** will be drawn), set of activated tracks and a temporal position for playback. The reader may wish to compare this to the distinction between the **Document**, **DocumentReader** and **DocumentWriter** abstractions presented in section 3.5.

## 6.1. ARCHITECTURE MODEL AND API ASSESSMENT

The emphasis of libmedia is instead on providing abstractions for the data elements themselves, while object instances that perform various transformation and processing tasks on media data play only secondary role. Additionally, it strictly separates between the two types of objects. The benefit of this approach is that it allows the programmer to focus on the processing intent and be wholly unconcerned with the technical realization of media processing behind the scenes: For a video capturing application, the programmer's *intent* is to move images received from a camera to a data store; in this case, the *technical realization* may involve receiving data frames that represent images in some format, transformations to convert the data into a format suitable for storage, as well as framing and writing the data to the storage. Take for example the simplest possible code sequence that captures a video stream from a camera and writes it into a container file:

```
/* choose a file format */
ref<const DocumentFormat> fmt=lookupDocumentFormat("mpegps");
/* create new document */
ref<Document> doc;
doc=openDocument("output.mpg", O_CREAT|O_WRONLY, docfmt);

/* create a video track inside the document, using default
parameters for size, frame rate, compression etc. */
ref<const VideoTrack> vtrack=doc->createTrack<VideoMedia>();
/* list of tracks containing just our single video track */
Document::TrackList tracklist(vtrack);

/* open write accessor into the file; start writing at
temporal position 0, open-ended */
ref<Renderer> render=doc->openWriter(tracklist, 0, INFINITY);
/* obtain handle to single opened track */
ref<VideoRenderer> vrender=render->select<VideoMedia>();

/* choose capture device by OS name */
ref<CaptureDevice> dev=V4LCaptureDevice::create("/dev/video0");
/* select video channel */
CaptureDevice::ChannelList chans(dev->select<VideoMediaType>());
/* open selected channels */
ref<Source> source=dev->openReader(chans);
/* obtain handle to single opened channel */
ref<VideoSource> vsource=source->select<VideoMedia>();

/* capture and store */
ref<VideoFragment> fragment;
do {
  /* first read will trigger capture if not started yet */
```

```
    fragment=vsource->getFragment();
    if (!fragment) break; /* end of stream? */
    vrenderer->render(fragment);
} while(fragment->end().t<60.0);

/* close renderer so pending images are flushed and the
sequence is finalized */
render->close();
```

Despite the simple look of the example, the above code truly handles all aspects of capture and storage: If the capture device is supplying data that is not in a format suitable for writing into the designated file (e.g. image size mismatch, color model mismatch), all required transformations are performed transparently. This may also include decompression/recompression of the data, additionally the file writer implementation is sufficiently smart to figure out when no recompression is required – in this case, the compressed data as produced by the camera may be written "verbatim" into the file.

It is important to understand the role the retained-mode processing paradigm plays in this example – in an immediate-mode processing framework, the commands in the inner capture loop:

```
    fragment=vsource->getFragment();
    vrenderer->render(fragment);
```

would first read an image from the source, transform it into a universally-understood intermediate format (such as an uncompressed rastered image using a specified RGB color model), and subsequently transform it again into the representation required for storage. This would in almost all cases be inefficient as several of the required transformations can be short-circuited: The retained-mode processing model employed by `libmedia` can dynamically detect and use these shortcuts.

Here as well it is illustrative to compare the terseness of the above code snippet with QuickTime: Code samples provided by Apple that perform equivalent services approach 1000 lines of code[4]. Most of these lines are "boilerplate" code that manually interfaces with the QuickTime decompression and compression services, individually applies any required transformations, and is concerned with repacking data between the different types of structures expected by individual QuickTime API calls. Also, unlike `libmedia`, QuickTime puts the burden on the programmer to detect whether decompression/recompression is required at all – if s/he wishes to avoid recompression if the data is already in the correct format, s/he must implement an alternate short-circuit execution path for this purpose.

---

[4]See the sample code provided at:
http://developer.apple.com/samplecode/CaptureAndCompressIPBMovie, fetched 2007-08-24

## 6.1. ARCHITECTURE MODEL AND API ASSESSMENT

Essentially, the lack of data abstractions in QuickTime forces the application writer to spell out and hard-code the *technical aspects* of media handling that `libmedia` can figure out automatically, and for which optimized processing paths are automatically provided through the **Renderer** concept. It might be argued that the explicit spelling out of each processing step in QuickTime provides the programmer with a desirable more detailed level of control – this is however doubtful as the processing steps to be taken are usually dictated by technical constraints, leaving little for programmers to "choose". For the rare cases where this assumption may not hold, the developer may always resort to supplying a custom **Renderer** implementation.

### 6.1.2.2 Media types

QuickTime is firmly rooted on the assumption of time- and space-discrete representations for audio and video media, (i.e. PCM audio and time-discrete video sequences of rastered images, see sections 1.3.2.1 and 1.3.4.1). In contrast, `libmedia` can architecturally support *any* computable representation of these media types (cf. section 1.3), including true time- and space-continuous audio representations, through the abstract base interfaces introduced in section 3.3; discretized representations are merely considered an (important) special case. This means that there are types of media that `libmedia` can consider e.g. "video" while QuickTime cannot – one such example are flash animations which are not necessarily time-discrete (and additionally also not space-discrete as flash uses a vector graphics model).

QuickTime partially circumvents this problem through its modular support for different media types (for example "flash" animations are a distinguished type of media, with its own supporting infrastructure in the form of media handler components and API functions). It is however worth noting that `libmedia` subsumes several distinct QuickTime media types into a single unifying video media concept.

### 6.1.2.3 Control flow model

The imperative programming model offered by both QuickTime and `libmedia` generally allows the programmer to retain in control of the application's control flow – control is passed from the application to the media framework only when services are required, there is little "autonomous" processing. Several of the real-time capturing and rendering components however need some service processing asynchronous to the application's control flow – the application must therefore (infrequently) call into the library to "lend" control flow to the multimedia framework for this processing to happen.

In `libmedia` this need is addressed in a generic fashion through the abstract callback registration interface discussed in section 5.3.1 – this interface is used by all active objects, and the application is responsible for supplying an implementation of this interface. The application is thus effectively decoupled from the internals of the media processing library: To the application, the kind of object that registers for notification through these interfaces does not matter – servicing of event notification callbacks through this interface will satisfy all needs of the library.

Moreover, the interface is explicitly designed to be thread-friendly – the callbacks to deliver notifications for which libmedia objects have registered an interest may be provided by a secondary (or even multiple) threads of execution. This provides even better decoupling as the control flow of the main application thread does not need to be interrupted by periodic servicing of notifications. The mechanism also allows delegation of time-critical operations to dedicated threads to improve real-time behavior[5].

In contrast, QuickTime addresses this need through a number of functions such as **MoviesTask**, **MCIdle** or **DataHTask** that must be called "periodically" to provide processing time to the respective QuickTime components[6]. The application programmer is expected to globally track which objects need to be tasked: For example, the programmer must generally issue **MoviesTask** calls to let QuickTime progress with playback of a particular movie (if it is not called, video and audio presentation will stall[7]). However, if the programmer also instantiates a *movie controller* object (providing more high-level control functions on movie playback), the function **MCIdle** must be called instead.

The approach taken by QuickTime leaves much to be desired in comparison to libmedia – effectively it introduces coupling of "local" uses of the media framework to the "global" execution flow: If for example a sub-module of a larger program makes use of *movie controller* objects, then this sub-module cannot be used transparently by the application as the main execution flow must be modified to include the **MCIdle** calls (or wrapper calls provided by the sub-module). It is also not possible to decouple processing from the main execution flow by creating separate threads in the corresponding sub-module: The QuickTime components are not thread-safe.

---

[5]The registration interfaces for event notification discussed in section 5.3.1 do presently not provide any support to let the requester communicate desired QoS (quality of service) requirements, so the architecture can currently not support any hard real-time guarantees. Nevertheless, future development into this direction was envisaged and taken into account during the design of the interfaces.

[6]If the application uses high-level objects such as **HIMovieView**, these will schedule appropriate timer callbacks with the user interface toolkit and handle the required tasking calls transparently – however, these high-level objects are limited in usefulness to very specific application scenarios such as simple media playback.

[7]The QuickTime documentation is vague about the call frequency and recommends "10 to 20 milliseconds", the API also provides further functions through which it can communicate the suitable points in time when it would like to receive calls from the application. Experiments revealed that QuickTime buffers up to half a second of audio to avoid drop-outs in case the application is not servicing QuickTime in a timely fashion. Video playback requires at least one call per image displayed.

## 6.1. ARCHITECTURE MODEL AND API ASSESSMENT

#### 6.1.2.4 Delegation of processing

The immediate-mode processing model of QuickTime suffers from difficulties in meaningfully delegating individual processing steps (see section 3.1.3). Consider for example the following sequence of processing steps:

1. Decode two images from compressed representations
2. Create a new image by blending the two previous images
3a. Display the resulting image on the screen *or*
3b. Compress the resulting image and store it into a file

Processing should ideally be performed "close" to the final destination of the complete image to avoid unnecessary communication: If step 3a is to be chosen, then the prefered candidate for processing would be the display system – if avenue 3b is taken then it is more desirable to keep processing in a system component that has close ties to the storage subsystem. In both cases, it is the *last* step of the processing pipeline that determines the ideal processing environment.

If the components providing the individual operations described in steps 1 through 3a/3b are however architecturally isolated, then the steps may only be delegated *individually*: The component performing the first step cannot know where the second operation is to be performed and must therefore unconditionally make the result of the processing operation accessible to the calling application (e.g. by communicating back the results).

While it is possible to extend an immediate processing API by providing 1a/1b and 2a/2b variants of the above processing steps (that do not differ semantically but delegate the operation to different processing units in accordance with 3a/3b), this only increases the burden on the application programmer: S/he must now decide from the very beginning where processing is to take place, and provide two different code paths dealing with the different forms of delegation.

On the other hand, the retained-mode processing concept employed by `libmedia` addresses this problem in an architecturally clean way: The full logical processing chain is established before any operation is executed, any decision to delegate the operations can take all parts of the chain into account. The concept is of course generic and allows delegation to many different processing entities: This includes networked display systems (see chapter 4), any types of co-processors (e.g. GPUs) or even distributed processing on more than one node.

### 6.1.3 Comparison to DirectShow

DirectShow as a prominent representative of filter graph media toolkits features a processing model that is very distinct: The application must construct a filter graph that represents the chain of processing steps to be executed for a stream of time-based media. Execution of the graph is autonomous and asynchronous to the rest of the application (it is automatically relegated to a different thread). The media data itself is not directly available to the controlling thread – it is only avaible to the nodes in the filter graph, therefore an application is forced to create its own filter component that will at some point in time passively receive the data.

The filter graph processing model is very popular in the field of multimedia as it is easily understood and provides a straight-forward model to develop and extend the architecture. In particular, the service of completing a graph through automatic introduction of auxiliary format conversion nodes provides a significant benefit over the simple and explicit model of QuickTime as it relieves the programmer from manually constructing the full processing pipeline. This benefit is however traded in for an execution model that forces the programmer to relinquish control and is significantly more restrictive.

#### 6.1.3.1 Data model

DirectShow and `libmedia` both allow to relieve the programmer from explicitly constructing the full media processing pipeline. While both achieve this goal through meta-data annotations of some sort, the level of abstraction provided for the data elements is very different. From DirectShow's point of view, the format annotations (**AM_MEDIA_TYPE**) attached to each data element is merely a "label": Its only purpose in the context of the architecture is to serve as a mechanism for matching filter pins. In particular, two different formats may describe a data element that can conceptually be understood as an "image" albeit in different representation formats (e.g. a compressed and a rastered representation), but the two labels identifying the different formats need not have any relationship to each other. The most the architecture can "know" about any two different formats is whether there is a filter capable of facilitating conversion between the two formats. Furthermore, the media samples exchanged between two filter nodes can usually not be interpreted out of the larger processing context: Each filter node may hold some implicit state (such as initialization data that has been passed through the graph during initialization, or previously decoded images used as reference pictures) that is required to usefully process a sample.

In contrast, each media element in `libmedia` carries all information (e.g. initialization data, decoding dependencies) to allow a correct interpretation of the contained data. Naturally, it would be wasteful to unconditionally discard any acquired state during rendering and start over for each individual fragment – **Renderer** implementations will keep some state in internal caches (see section 3.4.3.2). The distinction between the two approaches in practical terms is therefore that filter graphs implicitly hold *hard* and *irrecoverable* state required for *correctness*, while the state information by `libmedia` **Renderer** drivers is *soft*, always *recoverable* and merely kept for *efficiency*. This allows a more flexible processing model as the processing can safely be started, interrupted and resumed

## 6.1. ARCHITECTURE MODEL AND API ASSESSMENT

at will. This additional flexibility can be put to good use in several application scenarios.

#### 6.1.3.2 Dynamic optimization

Consider an application that wishes to process a video stream: Most of the stream is to be left unmodified, but a few images are to be modified, e.g. by adding graphical elements. In a filter graph environment this processing would be realized as a filter node that requires rastered images as input and produces rastered images as output: the filter can selectively modify images or pass them through unmodified.

The introduction of this filter into the processing graph has a number of consequences: Due to its input and output format constraints, auxiliary conversion filters may be introduced during graph building to coerce transformation of the image into a format suitable for drawing. This means that conversion is forced *even* if our drawing filter does not perform any operation at all.

Consider on the other hand the quite obvious realization using the libmedia API:

```
while(true) {
  /* retrieve image from source */
  ref<VideoFragment> fragment=source->getFragment();
  if (!fragment) break;

  /* modify images covering temporal range [60..65] */
  if (fragment.end()>=60.0 && fragment.begin()<=65.0) {
    ref<Image> orig=fragment->getImage();
    ref<PaintedImage> painted=PaintedImage::create(
      orig->width(), orig->height(), orig->aspectRatio());

    /* copy original image */
    cairo_pattern_t *pattern=orig->getCairoPattern();
    cairo_t *ctx=painted->getCairoContext();
    cairo_set_source(ctx, pattern);
    cairo_paint(ctx);
    cairo_pattern_destroy(pattern);

    /* draw black rectangle */
    cairo_rectangle(ctx, 50, 50, 200, 50);
    cairo_set_source_rgb(ctx, 0, 0, 0);
    cairo_fill(ctx);

    /* draw text in black rectangle */
    cairo_move_to(ctx, 75, 75);
    cairo_set_source_rgb(ctx, 1, 1, 1);
```

```
    cairo_show_text(ctx, "CENSORED");
    cairo_destroy(ctx);

    /* pack image into new fragment, reusing the timestamps */
    fragment=VideoImageFragment::create(
      fragment->begin(), fragment->end(), painted);
  }

  /* pass fragment to renderer */
  renderer->render(fragment);
}
```

In this case, most of the fragments are passed from the source to the renderer unmodified – only images in the temporal interval $[60; 65]$ are redrawn. This in turn allows the renderer to *dynamically* optimize the processing required for each fragment individually: When no drawing operation is to be performed, no superfluous transformations are performed. In extreme cases, both source and renderer expect the media elements to be in the same compressed representation format: This would result in decompression/recompression of only the retouched images, while the surrounding images are passed through verbatim[8].

Avoiding unnecessary transformations has two important consequences: Most obviously, the required processing time can be saved and used for other tasks. Additionally, this also reduces potential rounding or other numerical errors that could be introduced by imperfect implementations of conceptually mutually inverse operations. The impact of this optimization heavily depends on the use-case: For example, video editing often uses long scenes of source material with no or relatively minor modifications (such as color correction), while only a small number of images have to be retouched to achieve e.g. scene transition effects (see figure 2.7 and page 78 for an example of a realization using the `cairo` API).

Achieving the same benefits in DirectShow would require the programmer to construct separate filter graphs to represent processing with and without the additional drawing node – but the filter nodes also have the important role of hiding the intricacies of temporal compression through internal buffering and state management: Splitting processing into two graphs forces the programmer to handle this himself. Such an implementation would therefore require a considerable amount of fix-up code and requires a thorough understanding of the underlying formats.

The important point is that the `libmedia` API provides this service essentially for free: The most straight-forward way for the programmer to express the processing intent already enables these optimizations.

---

[8] Some of the surrounding images may have to be recompressed as well, see the next section for a discussion of temporal compression.

## 6.1. ARCHITECTURE MODEL AND API ASSESSMENT

### 6.1.3.3 Temporal compression

One of the explicit motivations for filter graph approaches is to accommodate the requirements of temporal audio and video compression: Samples corresponding to different points in time are not encoded independently but require varying degrees of "context" (see also the definitions 13 and 14 on pages 42 and 50 respectively) which is provided by the implicit state held in the filter nodes.

While this works well during processing of a "complete" sequence of media data (i.e. a sequence that does not have any references to outside frames), it just pushes the real problem out of the processing architecture itself: If an application does not want a complete sequence but just a small part of it (maybe down to a single image), how many frames of context must be fed into the graph so that the desired media elements can be extracted? The solution adopted by DirectShow and most other filter graph approaches cheats around this problem by providing a rather vague "best-effort" service guarantee: The decompressor filter nodes must accept incomplete sequences of frames, but on the other hand may not produce any output audio samples or images until they have accumulated sufficient state[9]. While this approach may be suitable for real-time playback of stored media (where the screen stays black or the audio remains silent for a short duration of time after a random seek, until a resynchronization point is reached), it is inacceptable for many other forms of media processing.

A similar but even more complicated problem arises during media compositing: Revisiting the example from the previous section, it would be desirable to only re-encode the minimum number of images – in the case of temporal compression this would not only be the modified images, but also "a few" temporally neighboring images that use the original images as reference. The filter graph does not allow a precise identification of these images: The required information is hidden inside the black box filter nodes.

The approach taken by `libmedia` does not suffer from any of these difficulties as all dependence information is explicit in every fragment. In particular, it supports both true random access to every point in time of stored media content as well as minimal recompression in case of storing processed sequences in a sufficiently similar format. Support for this service has been implemented for the MPEG-1/MPEG-2 video representation formats as a common example for complex temporal compression, but the realization is sufficiently generic that an implementation for other formats (e.g. H.264) is possible as well.

It should be noted that the concept of minimal re-encoding for temporal video compression techniques such as MPEG-2 is not particularly new, as many specialized applications can also perform this service – they are however stand-alone and not integrated with the existing media processing frameworks due to their

---

[9]This is particularly problematic for compressed representations that require some initialization that is encoded only once at the beginning of the compressed data stream – in the best case, the data source has sufficient understanding of the format that it can extract the initialization data and (re-)inject it during processing startup even if playback does not start from the beginning of the sequence. This of course severely bends the modularization concept of filter graph architectures that strives to keep individual filter implementations independent!

limitations. The novelty is that `libmedia` provides a clean concept for integrating such optimizations, while conveniently making these transparent to the programmer.

#### 6.1.3.4 Data transfer model

The implicit data flow model of DirectShow also results in other limitations in situations where an application would conceptually like to reconfigure the processing chain at run-time: A stored media container may contain multiple alternative video or audio tracks, representing e.g. different camera angles or different language versions of the same content. An application may wish to provide the capability to dynamically switch between the alternative tracks during playback.

This provides a difficult problem for DirectShow – the implementor may either choose to interrupt the playback for reconfiguration of the graph, since the architecture does not allow precise determination of suitable resynchronization points the disruption may last considerably longer than a single frame. Or, the implementor may create a graph that decodes *all* contained tracks speculatively, using a multiplex filter on the decompressed image and audio data to select those that should be presented. Both options are obviously not completely satisfying.

The same problem on the other hand is conceptually very easily solved using `libmedia`: The loop handing over data to the **MediaRenderer** just has to read fragments from a different **MediaSource** representing a different track – any decoding dependencies are transparently handled by the renderer, resulting in instantaneous and disruption-free switch-over.

#### 6.1.3.5 Document abstractions

The DirectShow architecture concentrates on the processing of media data sequences, providing the unified abstractions of "source" and "sink" filters that read from or write to media container files – their role is quite well comparable to the **DocumentReader** and **DocumentWriter** concepts of `libmedia`. However, DirectShow lacks any abstraction equivalent to the **Document** concept, it only provides accessors[10].

This leads to problems in realizing applications such as time-shifted playback and recording: An application would like to capture (multi)media content into a container file while at the same time playing back data previously written into the container. Ideally, playback could be paused and restarted at will without affecting the capturing process.

The lack of a **Document** model in DirectShow however makes this scenario problematic: Capture and playback must be placed into distinct filter graphs as they would otherwise be executed in lock-step (thus, stopping playback would

---

[10]DirectShow certainly provides utility classes that allow to access the contents of e.g. an AVI file, but but there is no unified media container file abstraction that is implemented by multiple classes.

## 6.1. ARCHITECTURE MODEL AND API ASSESSMENT

also stop capture). Creating separate filter graphs – one for capture using a writer into the file as sink, the other for playback using a reader into the file as source – leads to coordination problems, as there is no instance that can coordinate access into the same file. Supporting this scenario in DirectShow therefore requires the application to provide its own customized source/sink filters that coordinate their access into the same container – the built-in accessor components are useless in this case.

Again, the solution offered by libmedia in this particular scenario borders on the trivial: The separation of documents and accessor objects allows the required concurrency of reading and writing. Both playback and capture can be implemented independently in the usual way without any special consideration, using either separate threads or an event-driven model.

### 6.1.4 Limitations

While the preceding sections demonstrated several of the considerable benefits, it should not be neglected that there is one specific area where the model chosen for the architecture is known to be problematic: Audio or video compression schemes that allow decoding dependency chains of unbounded lengths (note that definition 14 on page 50 does not exclude this possibility).

While the architecture of libmedia certainly allows processing and playback of these type of video or audio sequences using all of the available mechanisms (including delegation of processing to the X server), a strict interpretation of the **CompressedAudioSignal**, **CompressedImage** (see section 3.3.3) and **Fragment** concepts would consume unbounded amounts of memory: each **Fragment** must be self-contained and allow reproduction of the contained media data element from scratch, therefore the complete dependency chain would have to be preserved just for the last fragment.

On the other hand there are perfectly valid applications that use this kind of media representation – one example is given in [51] where every image (except for the initial one) is encoded relative to the directly preceding one: A simple application that just wishes to display each image in sequence can operate with bounded memory as the data to every image besides the previous one can safely be discarded before processing the next image. However, libmedia cannot know of this intended access pattern as there is no mechanism to inform the library that parts of a sequence can be discarded (and it would be difficult to provide such an interface without compromising the overall architecture).

This dilemma is resolved by providing a mechanism to "cut back" overly long dependency chains with thresholds that can be chosen by the respective format handlers – e.g. as soon as the dependency chain of the current fragment reaches a threshold length (of, say, 30 images), an older image (say, the fifth-youngest) is decompressed and the dependency chain shortened accordingly (in this example resulting in 25 images that are not required any longer). This forced decompression obviously violates the "lazy evaluation" principle, although in a way that does not disrupt the architecture – the architectural mechanisms provided for caching of computed results have a similar result in practice. There is however the matter that cutting back the dependency chain requires a **CompressedImage** to be replaced by a **PixelImage** as reference frame for the fourth-youngest image – this clashes with the principle of providing *immutable* data elements and therefore requires a number of precautions.

While this solves the problem of unbounded memory, a naive implementation introduces a new inefficiency: Since the architecture allows delegation of decompression operations, the image might be decoded twice – once by delegation through a **Renderer** driver in a remote target device, later a second time by forced evaluation inside the application to allow cutting back the dependency chain. Therefore, a second mechanism exists that allows tracking of decompressed representations of images in remote devices, as well as provisions for fetching back in preference to duplicated processing. Note that this retrieval of the decompressed image from a remote location can also benefit from the "lazy evaluation" principle – i.e. the data is only fetched when the sampled representation is actually explicitly requested through the **sample** method.

The combination of the above two approaches manages to keep memory consumption bounded without introducing additional processing, but they require the existence of a bi-directional communication channel with the remote decompressor – for the purposes of correctness it does not matter if the channel is slow, but in the absence of such a channel, the library *must* duplicate processing. Note that no actual communication is performed in the common use-case of just decompressing the images for example on a remote display – here, `libmedia` would merely perform the required tracking to be certain that the images *could* be fetched back if needed.

In this scenario, an immediate-mode approach to media processing indeed has the advantage of simplicity – if the application exactly knows that none of the reference images will be reused in the future, it can safely discard them. While the `libmedia` architecture manages to achieve the same processing efficiency, it does so only through quite some trickery.

## Summary

The media architecture processing architecture presented in chapter 3 compares quite favorably with existing architectures such as QuickTime and DirectShow on a conceptual level. It provides an easy-to-use API, allowing application programmers to express their media processing intent in a fairly clear and terse fashion.

Compared to QuickTime, `libmedia` provides the advantage that the **Renderer** concept relieves the programmer from the technical aspects of media processing; where the QuickTime programmer would have to manually code most parts of the processing pipeline, `libmedia` automates this task using the rich meta-data attached to each media element. Additionally, `libmedia` also provides better support for delegation of processing.

Compared to DirectShow, `libmedia` provides the advantage of better dynamic optimization capabilities, in particular the **Fragment** concept and precise tracking of coding dependencies provides a high degree of flexibility to the application programmer that cannot be matched by the static processing model of DirectShow.

It should be noted that in all of the scenarios described in section 6.1.3, QuickTime would basically show similar efficiency advantages over DirectShow as `libmedia` does (but using considerably more lines of code).

## 6.2 Efficiency evaluation

The preceding sections pointed out the benefits of the dynamic optimization capabilities offered by the retained-mode execution model of `libmedia` from an application programmer's perspective. While several high-level optimizations enabled by this approach where already mentioned there, the discussion has so far remained on the conceptual level without any quantitative analysis of the costs and benefits – the purpose of this section will therefore be to provide the data to back up the assertions made.

All empirical results were obtained on a dual CPU Opteron 244 system running Debian Linux 4.0.

### 6.2.1 Overhead

The performance characteristics of immediate-mode multimedia processing can be understood fairly easily as each operation is executed synchronously to the control flow of the controlling application. Moreover, very little management overhead is incurred as the sequence of previously applied operations can be "forgotten" as soon as they are finished. In contrast, the retained-mode processing paradigm introduces overhead for bookkeeping as it needs to keep track of all state information associated with a media data element until the time it is needed. Further overhead is introduced by the renderer optimization framework (cf. section 3.4.3.2).

The following results give insight into the overhead introduced by the retained-mode model.

#### 6.2.1.1 Compressed media handling

The first experiment compares two different implementations of a classic "media player" application, once using the retained-mode API provided by `libmedia`, and once using a customized pipeline using the immediate-mode approach featured by traditional media processing frameworks. The second application was

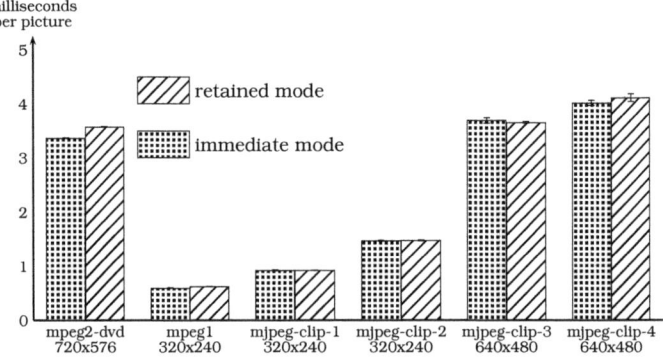

Six different video clips were processed using either the retained-mode interface offered by libmedia, or an immediate-mode processing pipeline customized for the particular task.

Figure 6.2: Comparison of retained- and immediate-mode processing of compressed media files

specifically optimized for the task of playing back *one* specific type of media data, using format-specific shortcuts where possible[11]. Both applications use the same decompressor implementations: While the retained-mode player uses the **CompressedImage** abstraction that sufficient meta-information to allow a **Renderer** to automatically instantiate a decompressor, the immediate-mode player bypasses the infrastructure provided by libmedia and calls directly into the decompressor instances.

This is intended as a "worst-case" comparison for the scenario of a simple media player, falling back to non-delegated software rendering inside the client application: No actual transformation is applied to the media data, therefore repacking of the data into **CompressedImage** and **Fragment** containers with semantic annotations is just overhead.

For the retained-mode processing case, the media data is stored in an AVI container and the ordinary **Document** and **DocumentReader** mechanisms are used to read the media as individual **Fragment**s from the file. Each fragment is then handed over to a custom **Renderer** implementation that just coerces all images into a rastered representation by simply calling the **sample** method and subsequently discards the data. For the immediate-mode processing, the data is already partitioned into individual frames so that processing essentially only consists of calls into the decompressor itself (the same decompressor implementation was used in both cases). As figure 6.2 shows, the experiment fails to

---

[11]For example, the custom pipeline merges reading from the data source and decompression into a single step. The libmedia **DocumentReader**s will instead locate and retrieve complete frames before they are returned to the caller who may pass it to a **Renderer** for decompression. Other micro-optimizations include pre-allocating all required memory.

## 6.2. EFFICIENCY EVALUATION

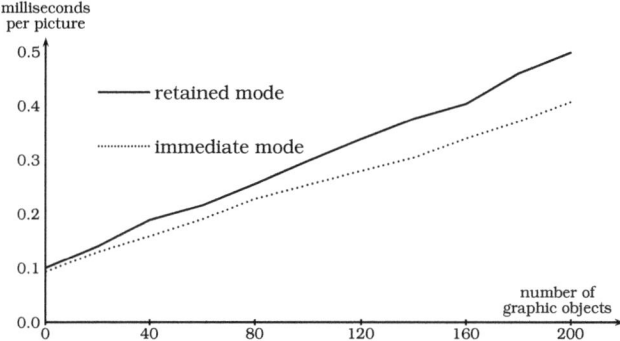

The diagram shows the time taken to construct an image using different numbers of graphic objects (colored rectangles in this experiment). While the retained-mode model exhibits a measurable overhead per object, the overhead is lower than the time taken to actually draw the object, and also low compared to the base cost of touching every pixel of the drawn image at least once. For low object counts the overhead is therefore vanishingly small.

Figure 6.3: Comparison of retained- and immediate-mode drawing

exhibit any significant overhead introduced through the various `libmedia` abstraction layers and its retained-mode processing model: Even though the measurements show excellent repeatability with very little variation, the difference is neglegible.

### 6.2.1.2 Compositing

While the previous experiment established that the bookkeeping overhead for compressed media representations is neglegible in comparison to the computations required to transform the compressed into a rastered representation, the next experiment will discuss compositing. Since compositing operations may be computationally less demanding, it is expected that any overhead will more easily show up in this scenario.

The second experiment is intended as a worst-case scenario for image drawing: Compositing of a full image has to touch each pixel at least once, but the numbers of pixels affected per each operation may vary. Figure 6.3 shows the time required to create a $320 \times 240$ image, using either retained- or immediate-mode processing. The image creation process involves filling the canvas using a specified number of colored rectangles (the rectangles were chosen disjoint so each pixel is painted exactly once). As the diagram shows, the relative overhead induced by the construction of the **PaintedImage** intermediate representation used for retained-mode rendering does indeed grow with the number of objects drawn – the overhead is however low compared to the cost of the actual drawing operation. Additionally, this must also be put into perspective to the number of compositing operations required to achieve typical effects: For example, the blend transition shown in figure 2.7 on page 78 can be expressed using only three compositing operations.

The overhead should also be considered in relation to the timing results given in the previous section (note the order of magnitude difference in the temporal scale between figures 6.3 and 6.2): In the presence of media compression or decompression operations (or other similarly expensive operations), the overhead incurred for retained-mode compositing is still lost in the noise.

#### 6.2.1.3 Media playback on remote X displays

For the next experiment, a video player application was started on a remote system, with the graphical user interface redirected to the local system. The interaction of the application with the display system used the X protocol through an established TCP connection over a standard fast ethernet physical layer.

For illustration purposes, two different types of players were used:

- The ubiquitous `vlc` player, which does not have any support for the COMPRESS extension and therefore must resort to transmitting uncompressed images through the X protocol.

- The simple media player mentioned in section 6.1.1.1 that uses the X extensions from chapter 4 through the **X11RendererDriver**.

The content for playback was chosen to be a standard commercial DVD, containing MPEG-2 encoded video at an average bitrate of about $4.8$ MBit/s.

`vlc` fared poorly in this experiment, completely saturating the 100 Mbit/s network link while playback was still jerky. The `libmedia`-based player on the other hand was able to provide smooth playback transmitting about 5 Mbit/s of data (audio was suppressed for easier comparison as `vlc` does not support networked audio).

The X protocol interactions of both programs and the server were subsequently analyzed using `tcpdump` and a small improvised protocol decoding utility. For `vlc` the analysis revealed that it uses the XVideo extension to upload images in a $Y'C_b'Y'C_r'$ format[12] to the server, thus would require $720 \times 576 \times 25 \times 16$ bits per second (or about $158$ Mbit/s) for the image data alone. For the `libmedia` player, the following pattern of X protocol requests was generated for each image to be displayed (see also figure 4.7 on page 142):

- `CreateFrame`, containing the XID of the server-side **ImageDecompressor** object, the ID of the frame to be allocated as well as between zero and two IDs for reference frames. The encoded request is 12, 16 or 20 bytes in size, depending on the number of reference frames used.

---

[12]The format stores one $Y'$ luma value per each pixel and one $C_b$ and $C_r$ value for each second pixel (cf. section 1.3.3.2). The values are stored interleaved in the given order.

## 6.2. EFFICIENCY EVALUATION

- `FrameData`, containing the XID of the server-side **ImageDecompressor** object, the ID of the frame for which data will be submitted, as well as the frame data itself and a length specifier. The encoded request is 16 bytes in size, plus the size of the compressed image.

- `Schedule`, to issue scheduled operations for later execution in the server: Parameters to the request identify the scheduler, the points in time for execution and expiry of the enclosed group of requests, as well as flags requesting notification about execution or expiry of the group. The request contains the following "piggy-backed" requests:

    - `Decompress`, containing the XID of the server-side **ImageDecompressor** object, the ID of the frame to be decompressed as well as the XID of the "picture" resource to be associated with the decompressed image. The encoded request is 16 bytes in size.

    - `RenderChangePicture` to set several parameters on the decompressed picture structure identified through its XID (in particular, chooses bilinear interpolation mode for scaling). The encoded request is 20 bytes in size.

    - `RenderSetPictureTransform` to set up an image scaling matrix on the decompressed image, so that it is horizontally stretched on subsequent blit operations to correct for the aspect ratio assumed by the MPEG-2 video. The encoded request is 44 bytes in size.

    - `RenderComposite` to transfer the decompressed image into a backing pixmap for the window where the video is to be shown on screen. The encoded request contains the XIDs of source and destination pictures as well as various parameters to identify their relative position, clip rectangles and the compositing operator to be used for the transfer operation (in this case the OVER operator). It is 36 bytes in size.

    - Another `RenderComposite` request to transfer the picture from the backing pixmap to the visible frame buffer. The encoded request is again 36 bytes in size[13].

    In total, this request is 176 bytes in size.

- (At a later point in time): `ReleaseFrame`, allowing the server to discard all data associated with the given frame ID. The request is 12 bytes in size.

The server honors the request for notification (send in the `Schedule`) request by generating a completion event for each group of scheduled requests (each X event is encoded through 32 bytes). The overhead incurred by the X protocol

---

[13]The indirection through the backing pixmap is not strictly needed, the renderer driver could blit directly to the visible frame buffer. However, copying the data into a backing store allows the server to recreate the window's content if needed without requiring a round-trip to the client to request redrawing – this would pose great difficulties in the case scheduled drawing, as the client cannot know with sufficient precision which image is supposed to be visible currently.

# CHAPTER 6. ASSESSMENT

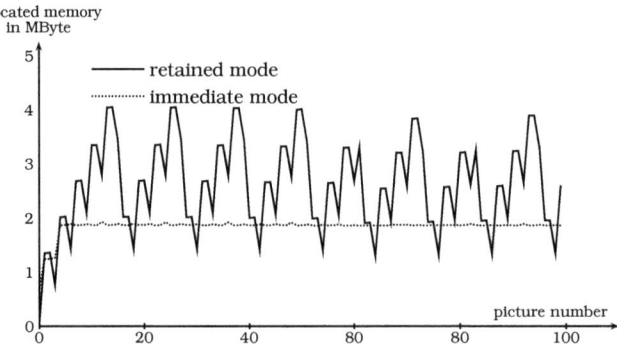

*The graph above shows the total amount of allocated memory during processing of an MPEG-2 video sequence consisting of a mixture of I-, P- and B-frames. The memory usage snapshots were taken at the time the numbered image corresponding to the x coordinate was displayed. Notice that the allocated memory totals depicted in the graph also include the memory required to hold the compressed data. In fact, the slightly largere compressed I frames are clearly visible as small "bumps" in the dotted line.*

Figure 6.4: Memory usage during processing of an MPEG-2 video sequence

interactions thus add up to about 224+32 bytes per frame, or 6.4 Kbyte/s. In the scenario at hand this is about 1% of the nominal data rate required for the compressed images.

#### 6.2.1.4 Memory usage

The processing concept of `libmedia` requires that any media element is self-contained and can at the very least be converted into a sampled representation. This means that all data required for reconstruction of a media element must be retained up to the point where the element is no longer reachable by any caller. For media representation formats that use predictive coding this retention policy transitively extends along the whole dependency chain.

At the same time the architecture makes extensive use of caching to avoid repeating transformations the result of which has already been computed. While cached data could in principle be discarded under memory pressure, the implementation is by default very conservative and attempts to keep cached results until they cannot be reached any longer.

The combination of the two concepts above means that considerable amounts of memory could in theory be required (section 6.1.4 already discussed a pathological case that might without further precautions lead to unbounded memory usage) if the dependency chains grow sufficiently long. This is however rarely the case in practice where the lengths of the chains are almost always bounded

## 6.2. EFFICIENCY EVALUATION

to relatively low values. Figure 6.4 illustrates the memory consumption of two media player applications, both playing back an MPEG-2 video stream from a commercial DVD [62]. The first employs an immediate-mode approach to processing and uses just the bare minimum of memory to hold at most three images concurrently. The retained-mode media player on the other hand was created using the standard libmedia interfaces. The figure makes the dynamic behavior of the retained-mode player clearly visible: Memory consumption due to cached images grows as the length of the dependency chain grows, but collapses regularly as soon as intra-coded images are encountered in the stream.

The behavior exhibited in this scenario is fairly typical for practical applications – libmedia has a higher memory footprint than immediate-mode processing toolkits, though not to an extent that is dramatic or would severely restrict the usefulness of the framework. Although there is some room for improvement, the higher memory consumption is to some extent inevitable by the design as the automatic tracking of resources must be conservative to ensure each reachable media element can be reconstructed at any point in time. The phenomenon is comparable to the situation in many other frameworks using some form of garbage collection (see e.g. [25]).

### 6.2.2 Audio latency and the X Window System

While the previous sections considered the characteristics of the libmedia framework, the new X extensions introduced in chapter 4 are useful in their own right, and merit some investigation outside the context of the larger multimedia framework. In particular, the AUDIO extension (see section 4.1.2) introduces completely new concepts (for processing audio) which the X server originally has not been designed for. The purpose of this section is to investigate how the X.org reference implementation of the X server fares under these circumstances:

Many audio applications are latency-sensitive, and the design and implementation of this extension tries to take this into account through various measures. In particular, the shared memory mechanism (see sections 4.1.2.1 and appendix B.2) helps maintain low latency for applications running on the same physical machine – since remote displays incur some additional penalty in terms of latency due to the required network communication, they are per se unsuitable for many highly latency-sensitive applications.

The implementation achieves this mostly through a good decoupling from the computationally expensive graphics operations also performed by the X server (see appendix B.1). It has however turned out that the X server implementation used for this project (the X.org reference server) has several latency issues – these are not limited to audio only, but are much more noticeable than with graphics. Structurally, the server is single-threaded and uses an event-driven model to dispatch client requests. As a result, any delay incurred by compute-intensive graphics operations affects the processing of requests of all other clients, in several scenarios a single client is able to completely monopolize the server's pro-

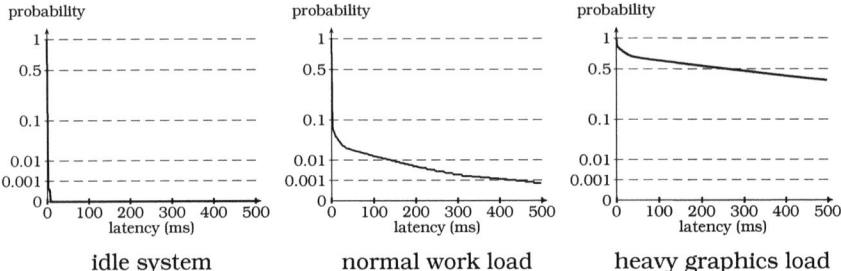

Figure 6.5: Latency characteristics of the X server

cessing resources. The issue becomes particularly visible (or maybe *audible*) for relatively simple requests that affect large amounts of graphics state, such as moving a window.

To illustrate the ensuing problem, a simple player application was created that uses either shared memory or the X protocol as transport mechanism for audio sample data. Both players need to synchronize the rate at which they supply data to the audio device's clock. This is done by using the TIME extension to request periodic "timestamp" events (cf. section 4.2.3) through the X protocol. If the X server is busily performing any graphics operations, these timestamp events may be delayed, as protocol processing is performed in the same thread[14] – figure 6.5 illustrates that even relatively trivial operations can easily have disruptive effects on the worst-case latency.

The unpredictable latency poses a problem for audio as applications must compensate with sufficient buffering to overcome latency spikes. This buffering however increases to the end-to-end latency between the application and the audio device. This ill effect of buffering can be mitigated through speculation, where the likely operation is scheduled way ahead of time but revoked in case of unforeseen events (see section 5.2). As the graphs show however, the amount of buffering and speculation required can quickly become ridiculous. Furthermore, these mechanisms are designed to deal with infrequent latency spikes induced by packet networks – while they *may* be used to cover up for architectural deficiencies in the X server implementation, this is hardly a good idea. It would therefore be highly desirable to improve the quite bad latency behavior of the X.org reference server implementation (see section 6.3.2).

Note that the latencies shown in figure 6.5 only affect the X protocol itself, but a player using shared memory can bypass the protocol entirely at least for data

---

[14]Of course, all other parts of the protocol processing such as request dispatching are affected by this delay as well.

transport. During all of the experiments, a player application running on the same system as the X server was therefore capable of providing uninterrupted playback using a relatively small buffer of less than 10 milliseconds (this also shows that the latency problems are really with the X server's architecture itself and not any of the other system components). But synchronizing with the server's playback rate becomes more difficult as the information provided by the timestamps becomes heavily "distorted" through the delays.

## 6.3 Future work

Since the creation of a full-featured implementation is an enormous task, the implementation provided in this project can – despite its conceptual merits – not compete with other frameworks in terms of the number of functional components e.g. for decompression. Consequently, there are several avenues for future developments:

- Since the architecture is kept highly modular, the first and most obvious direction for development is to provide new components that extend the functional capabilities of the implementation. This may include support for new input or output devices, storage or compressed representation formats for existing types of media, or even entirely new types of media. Since there is generally very little architectural work involved, this avenue may not be particularly rewarding from a scientific point of view, but it is of high importance for most practical purposes.

    One related area would be to implement support for the import and export of media formats like *flash* animations that are better described through the "compositing" facilities than the compressed media representation mechanisms (in the sense of section 3.3.3). It is expected that there are no major technical obstacles as the `cairo` API is all by itself already sufficiently powerful to express vector graphics[15].

- The *renderer* concept essentially acts as a sort of retargetable "compiler" that translates the given abstract media elements into a sequence of steps suitable for execution by the "target system" represented by the renderer driver. The concept also includes the possibility of optimizing the sequence of steps with respect to various metrics, such as minimizing computational effort or communication. The various optimizations implemented and discussed in section 3.4.3.2 are already sufficient to deliver the benefits illustrated in section 6.1, but the concept can of course be expanded further than has been possible within the constraints of this work. Several of the possibilities are outlined in more detail in the section 6.3.1 below.

---

[15]A "complete" implementation would however require a close investigation of how to map e.g. some of the more esoteric compositing operators supported by flash to `cairo`, possibly extending the set of supported operators.

- The implementation also uncovered several shortcomings in the existing infrastructure upon which the project was built. In particular, the X Window System is used in a way that is novel and (quite likely) not envisaged by the original designers. Furthermore, the X Window System was only used as an "output" device for media data, however it may also be useful to provide e.g. media capturing capabilities through the X protocol. There are currently several architectural obstacles that need to be addressed before this can be realized.

  Section 6.3.2 will discuss some of the deficiencies and how they could be addressed in future projects.

- The core media processing architecture is not very tightly coupled to the Linux environment chosen as target for this project. In fact, there is only one system-specific dependency for run-time listing and resolution of symbols (see appendix A.4) required for the modularization model (see section 3.2.1). Equivalent services are also available on other platforms, so that there are no obstacles to porting the framework to other targets. This also requires implementations of **Renderer** drivers that interface with the respective graphics system. Thus the implementation could be developed into a platform-independent media processing framework.

### 6.3.1 Future development of the "renderer" concept

Renderer drivers provide the interface used by `libmedia` to execute media processing operations for passing media to a designated "target", be that an output device or a storage container. The currently implemented renderer drivers already apply a number of optimization steps to reduce computational effort and communication, but this could be expanded vastly.

The implemented optimizations are a collection of relatively schematic transformations: They are unconditionally applied to given media elements whenever a matching transformation rule is found, so they are essentially "just" peep-hole optimizations. In typical applications this is sufficient as the representations of media elements are rarely more complex than expressions of three or less "irreducible" elements, so there are few (if any) optimization opportunities. The irreducibility is however a consequence of the optimization framework's inability to further "look into" compressed representation formats.

The case can be made for making the compressed formats less opaque than they presently are, as the current abstraction level precludes some potentially useful optimizations: Consider for example when transformations between two compressed media representations that store encoded DCT coefficients. The current implementation forces a transformation of the given data into an intermediate rastered image (thereby inverting the DCT to convert the coefficients into pixel values), and generating a new representation in the target format from this rastered image (thereby computing DCT coefficients from the pixel data again).

## 6.3. FUTURE WORK

It would be desirable to short-circuit this transformation by retrieving and encoding the DCT coefficients directly.

One possible avenue is to simply provide a "richer" interface to compressed formats that allows to extract various more abstract representations (e.g. in the example above, the interface could provide for a "block-DCT transformed rastered image" format). This approach was also suggested by other researchers [26], and some preliminary work towards this goal has already found its way into the current implementation by structuring decompressors and compressors accordingly. It is however unclear whether this leads to a generalizable concept.

A more radical approach would be to expand on the concept of JIT code generation already used for the specific sub-task of conversion between different pixel formats and color models (see appendix A.5). In the extreme, this could mean implementing media processing entirely in an abstract virtual machine and attempting to apply global optimization on a processing chain thus represented – but the experiences hitherto with JIT generation of code from VM byte code representations leave the author entirely unconvinced that this approach is sustainable (since none of the JIT optimizers encountered by the author were capable of matching the performance expectations). However, an approach using more high-level building blocks (such as block DCT, parsing of variable length codes from bitstreams) as equivalents of virtual machine "instructions" could prove feasible.

This would also be in spirit with the general architectural focus on describing the abstract representation a media element is coded in: Compressed media formats would in this case be given as an abstract definition of how the individual bits of the representation are to be interpreted as "media data". As a concrete example, instead of wrapping up an implementation of a Huffman decompressor and subsequent discrete cosine transformation into an opaque component, the format would consist of "declarations" that describe the use of these two algorithms and their relationship. The framework would then substitute suitable implementations, possibly generated on the fly and optimized to the provided parameters. This offers several new opportunities, especially with respect to delegation of media data transformations to co-processors: Code for the given target execution platform could also be generated just in time without writing .

### 6.3.2 X server infrastructure

For this project, the X.org reference implementation of the X Window System was used as most deployed implementations are based on it. The X.org architecture is quite flexible, allowing the implementation of the extensions required for this project without modifications to the core system. There are however a number of general architectural issues with the current implementation of the X server which the provided extensions happen to highlight.

As section 6.2.2 demonstrated, there are severe issues with respect to (predictable) latency behavior of the X server under load. The main reason for this is that the server is single-threaded and uses an event-driven model to dispatch individual client requests. As a consequence, long-running operations cannot be preempted, leading to conceptually unbounded latency.

Future work should address this issue by providing a multi-threaded dispatching model, either as dedicated threads per client connection or using a thread-pool approach. The required effort should however not be underestimated as it requires refactoring a large existing code base firmly built on the assumption of single threaded execution[16].

A second issue with the current X server implementation is that messages sent from the server to the client use a "semi-synchronous" interface: If the message is sufficiently small that it can fit into a single atomic message of the underlying transport protocol, then it will be sent asynchronously. If on the other hand a single message or the total of multiple messages is too large, then the server may block until the last part of the message has been transmitted to the client. This design is based on the assumption that there is rarely a need to transfer large amounts of data back to X clients. So far this has been justified: Applications that need to transfer image data from the server to the client are rare and thus do not provide a significant performance problem (and there is no other type of data besides images that could usefully be transferred).

If the X protocol were to be used as a more bidirectional communication mechanism, this would invalidate the underlying design assumption and thus provide a significant performance problem (this has also been verified experimentally – continuous transfer of bulk data through the X protocol even through local IPC mechanisms lead to "jerky" behavior of the X server and basically make the graphical interface unusable). As a prerequisite for providing capture of media data through the X server this issue would have to be addressed by reworking the I/O system of the X.org server implementation. In principle, this can be done within the constraints of the single-threaded event-driven model currently used by the server, but it is more useful to address this in the larger context of switching the server to a multi-threaded dispatching model.

## 6.4 Conclusions

This work presents the architecture of a new media processing framework developed for the Linux operating system environment (chapter 3). The architecture is complemented by the novel infrastructure introduced into the X Window System for delegated processing of media data (see chapter 4). Additionally, the issue of integrating the media framework as well as the X infrastructure with the rest of the software stack was considered in chapter 5.

The architecture provides a full-fledged multimedia framework, including capabilities to handle compressed data, media container files, capture and play-

---

[16]The fact that existing projects such as [67] only address a small portion of the overall problem, but nevertheless face severe difficulties should be seen as indicative of the feat to be accomplished.

## 6.4. CONCLUSIONS

back devices as well as very powerful media processing primitives from which complex video and audio effects like scene transitions or mixing can be constructed. The media architecture introduces several novel concepts that are not found in previously existing media frameworks (see chapter 2), such as *retained-mode processing*, *lazy evaluation* and weakly-typed media elements with *implicit format conversions*. These lead to a processing model that provides both a high degree of control, but at the same time enormous convenience: As section 6.1 illustrates, the model improves in several ways over previous approaches, by both providing an easy-to-use interface that relieves the programmer of many common and repetitive tasks, while at the same avoiding to force the programmer into a specific processing model.

The architecture is centered on the idea of building up an internal intermediate model, representing the image, audio signal, video sequence or other media element that the application programmer wishes to create. This model is retained, and can be interpreted by one of the various *renderer* drivers that form the end of the processing chain: Their role is to translate the model into a sequence of steps that can be executed on a specific target. The rich data model of the framework provides the renderer drivers with sufficient information to perform very powerful high-level optimizations that are basically inaccessible to other multimedia framework. All of these services are automatic, allowing the application developer to concentrate entirely on the the processing *intent*, while the framework determines a suitable and well-optimized technical *realization*.

While building up the internal model requires considerably more bookkeeping than more traditional immediate-mode processing architectures, the overhead is negligible and does not compromise efficiency (cf. section 6.2.1). In fact, weighing the sophisticated high-level optimization capabilities against the overhead leaves a substantial net efficiency gain.

What sets this framework even further apart from all previous work is the architecturally clean way in which it provides the capability for delegation of processing. In particular, this project addresses the long-standing issue of network transparency: While the X Window System provides a powerful networked graphics system, none of the previous projects have even attempted to extend this concept into the world of multimedia. In contrast, the framework presented here allows to delegate multimedia processing to remote display systems, using both network and processing resources sparingly. On top of that, the concept is not at all limited to mere media playback: Previously unthinkable applications such as network-transparent interactive video editors are enabled by this project.

The author therefore views these encouraging results as validation of the overall approach.

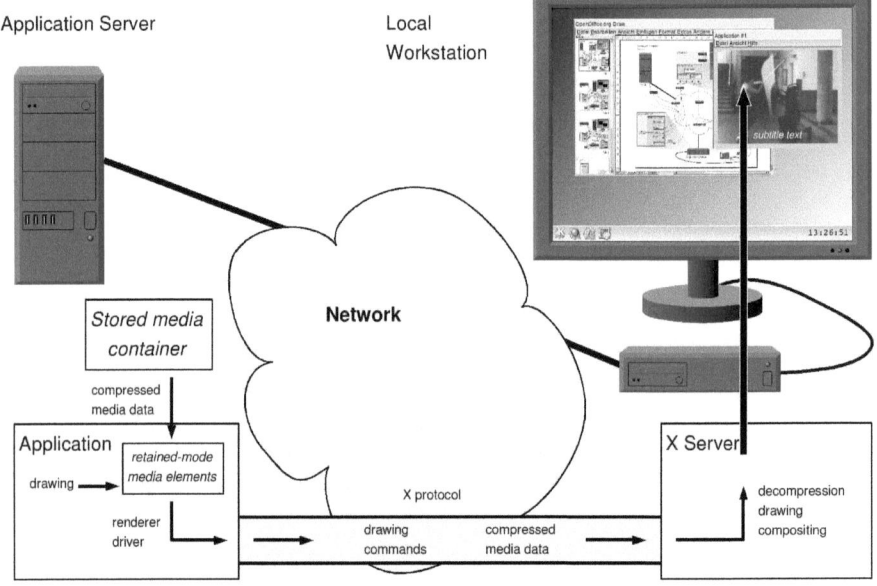

# Appendix A

# Implementation notes: Media processing library

This appendix briefly documents some notes on the implementation of the `libmedia` media processing library, the architecture of which was outlined in chapter 3. The notes concentrate on rather technical aspects of the architecture that are not dictated by functional requirements and that were largely discussed in chapter 3. Instead, most of the concepts that will be briefly touched here are dictated by performance requirements, interactions with the target run-time environment, or internal state management issues. Thus they are not part of a programming interface usable for writing media applications, but provide internal services usable for implementation of extension components.

The presentation is by no means complete and covers only few of the most important concepts – at 35000 lines of code an exhaustive discussion would easily exceed the scope of this entire work. The library has been implemented in C++ for a Linux run-time environment. The platform-specific portions of the implementation relate to the following areas:

- Operating system services (I/O events, threads, coordination and synchronization)

- Processor-specific services (run-time code generation)

- Execution environment services (dynamic linking, introspection)

- Device driver services (audio/video input/output)

These parts are well isolated from the platform-independent parts (e.g. operating system services that are not universally available through the same interface (like memory allocation, simple file and network I/O) are completely encapsulated in a separate auxiliary library that supports a wide range of targets), so that a port of the implementation to other target platforms is straight-forward.

All library services have been implemented in the namespace **media** to avoid naming collisions with other frameworks.

196  APPENDIX A. IMPLEMENTATION NOTES: MEDIA PROCESSING LIBRARY

## A.1  Data model

The implementation is designed to use available resources sparingly and thus tries to reuse any resources once they have been obtained. In particular, this means that data is passed around by reference instead of copying it, file handles are retained if they are required for future accesses, re-reading of data from network connections and files is avoided through semantics-aware caching[1] etc.

To support passing resources by reference the architecture generally uses two layers of objects to represent a piece of data:

- The *physical* layer: a specific resource such as a memory area, a file descriptor etc.

- The *semantic* layer: a media object such as an image, a media document etc.

Objects that belong to the physical layer are assumed to have *ownership* of the resource they represent – they act as *tracker objects* that keeps book of whether the resource is in use. Objects of the semantic layer in turn reference objects of the physical layer to hold the resources they require on their behalf. This distinction allows to reinterpret resources multiple times over their lifetime – for example, a **MemoryArea** object tracks a block of memory, which may pass through the following stages:

- First, the memory is allocated to hold data read from a file. The block of memory is then enqueued into a **BufferWindow** (see appendix A.3) for processing as part of a larger logical stream.

- A stream demultiplexer (see appendix A.2) inspects the data and assigns it to multiple streams. Our memory block may – as a whole or in part – be assigned to one of these streams, let's assume a video.

- Next, a parser (see section 3.3.3.1) may inspect the logical video stream and partition it into individual frames. The block – as a whole or in part – may be assigned to a **CompressedFrame** object.

- Alternatively, the block could also be assigned to a **PixelImage** object if the underlying format stores uncompressed images.

All of the above passing of the data through different processing layers is achieved without copying. The distinction of resource ownership and data interpretation also provides a clean way for applications to hand over application-managed resources to the library and cooperatively use them – this concept was already briefly mentioned in section 5.3.1.

---

[1] While actual media data is not suitable for caching (as it is usually processed exactly once), documents also contain meta-data (track information, indices) that are typically accessed quite frequently during the lifetime of an accessor to the document. The implementation is aware of this difference and treats the data accordingly.

While the effects are entirely beneficial in terms of resource consumption (and thus performance) it results in complex object lifetime rules. As has become clear during early development stages already, manual tracking of object lifetime is too cumbersome and error-prone to be feasible. Instead, a garbage collection mechanism has been implemented that automatically tracks the lifetime of all objects within the library. Since the C++ standard does not provide garbage collection as a language feature[2], a simple reference-counted approach has been taken. This scheme is viable for the case at hand because most data structures occuring can be expressed as acyclic directed graphs (and in the few cases where cycles are necessary for technical reasons, they have been realized using a "weak pointer" concept).

## A.2 Stream Demultiplexing

Many commonly used media formats (e.g. MPEG-1/MPEG-2 elementary, system and program streams; see [32] and [69]) do not provide an index table to locate individual elements of the coded stream (such as frames) to be looked up. Instead, the stream must be processed incrementally, locating the next element by either marker bits or length tags on the previous element. While this approach is complex, it has undeniable performance benefits and should therefore be prefered even *if* an index table is available: It allows to communicate the sequential data access pattern to the operating system which can then adapt and schedule read-ahead/drop-behind. Accordingly, this strategy is used where applicable, e.g. for properly interleaved AVI files.

Since the framework is designed to provide two different types of I/O semantics (see section 3.2.2) this means that stream demultiplexers must also be able to operate in two different modes:

- Blocking data reads and "pull" processing: One of the demultiplexed streams is pulled for new data, which may in turn call the demultiplexer for further data.

- Non-blocking data reads and "push" processing: The data source supplies new data to the demultiplexer, which distributes it to the individual streams who must then notify interested parties.

The state management for demultiplexing is further complicated by the fact that data consumers may mix both models freely – in particular, it is very common to request data in blocking mode during intialization to fill up any buffers, and switch to non-blocking mode afterwards. Furthermore, due to the referencing rules explained in the previous section, the object instance representing the stream demultiplexer and the object instance representing the demultiplexed stream may not hold references to each other.

---

[2]But see proposals such as [24].

It quickly turned out that a correct solution to this problem involves number of corner cases and is quite non-trivial. Therefore, libmedia provides a base **StreamDemuxer** class which provides the base for stream demultiplexer implementaions – it takes over all I/O operations and call processing for both the "push" and "pull" data models, and calls into a customizable function for actual stream parsing.

As a consequence there are a number of restrictions on how the stream parser must be structured: The stream parser cannot block and request more data for processing – instead, it must return to the calling **StreamDemuxer** indicating whether the currently available data permitted any parsing progress (which may or may not cause the **StreamDemuxer** to read new data and try again). The parser must therefore support incremental processing and must be structured as a state machine.

For many multiplex formats (such as MPEG-1/MPEG-2) the state machine required for processing is rather small (it consists of merely two states) – in this case the above limitations imposed by **StreamDemuxer** are not actually felt as such. However, more complex formats (such as AVI) involve four major states (not counting those required for error recovery) and moreover must support restoring states to (re)start processing from arbitrary points in the file.

Despite the complexity involved in the design approach chosen has shown its merit – in particular, it is to the author's knowledge the only implementation capable of supporting both "push" and "pull" processing for some of the more complex stream multiplex formats.

## A.3  BufferWindow concept

Since simplistic byte-level stream processing incurs considerable overhead, it is preferable to process larger chunks of data with random-access semantics instead. libmedia supports an efficient implementation of this common access pattern through **BufferWindow**: This class represents a small window into a conceptually infinitely large sequence of bytes. The window is delimited by "head" and "tail" pointers, and the data in between these pointers is held in a number of (not necessarily contiguous) buffers. Every byte within the window is randomly accessible, and **BufferWindow** supports operations that narrow or widen the window into both directions (note some conceptual similarities with the sample buffer concept introduced in 4.1.2.1).

## A.4  Dynamic symbol lookup

The library uses sub-namespaces to group factory objects providing a common interface together (cf. section 3.4). For example, the namespace **media::fileformats** contains the members **avi**, **quicktime** and **mpegps** which support instantiation of document objects of their respective types. The set of sub-namespaces is not predefined, in fact components may provide new sub-namespaces to facilitate dynamic registration (e.g. AVI file accessors use the namespace **media::avi::mediahandlers** to locate media handlers by their corresponding "fourcc" codes; any particular media handler provider may register an entry in this namespace to advertise their capability of functioning as an AVI media handler).

The library provides two mechanisms to access these namespaces:

- bind to a specific (named) symbol
- browse the list of all symbols in a sub-namespace

Generally, all symbols in the current executable (and all currently loaded libraries) are considered. In addition to these the library also considers a set of shared object files as potential sources for requested symbols (browsing includes their symbols and binding to a specific symbol causes the library to be demand-loaded transparently)[3].

Binding to symbols is achieved through the **dlsym**[4] call (possibly predated by **dlopen** to demand-load a library). Thus binding honors the usual symbol precedence rules (and e.g. allows an application to override functionality). Browsing requires access to the symbol tables of the current executable, but there is unfortunately no standardized and portable interface to the link editor maintaining these tables. On the Linux/ELF target used for this implementation the function **dl_iterate_phdr** allows to access the ELF section headers of all objects loaded into the current executable; the desired information can then be obtained from the DYNAMIC section (see [38]).

## A.5 Pixel format and color space conversion

Section 1.3.3 introduced several different color models, and section 1.3.4.1 introduced the concept of rastered images. In practical applications there is further variety as not all values of the color triplets are stored for every grid point – for example, it is for $Y'C_rC_b$ color models customary to provide $C_r$ and $C_b$ sample values with only half or qarter the resolution of $Y'$ samples (called *subsampling*; see section 1.3.3.2 for an explanation of the rationale). Other examples are digital cameras where it is not uncommon to sample "green" color values at double the resolution of "red" and "blue". Furthermore, the individual values may be either stored "planar" in separate two-dimensional arrays or "packed" together in a single array, using different ordering and different numbers of bits per component.

---

[3]In practice, this is realized with the help of a cache file that contains all exported symbols of a set of shared object files and indicates which file contains the symbol. Technically, this file is not necessary as all required information could as well be obtained from the shared object files, but its presence speeds up browsing and binding.

[4]POSIX.1-2001

## 200  APPENDIX A. IMPLEMENTATION NOTES: MEDIA PROCESSING LIBRARY

Such a definition as to where and how the sample values are stored is referred to as a *pixel format*.

Overall, this results in about 20 different combinations of pixel formats and color models that are used relatively frequently, and which practical applications must be prepared to handle. The processing paradigm of `libmedia` demands that an image using any of these formats must be convertible into any of the other. Direct conversion between the formats would require $20 \times 19$ transformation routines, using chained transformations through intermediate formats can reduce this number but may loose performance or precision or both.

The approach taken for this project uses just-in-time generation of code for transformation between two arbitrary formats. Foremost, this requires representing pixel formats and color models as data structures. The pixel format description partitions the image into elementary "macropixels" – these are the smallest repeating unit, taking into account subsampling of the color channels. The format description then specifies how sample values contained in one macropixel are represented as bitstrings, and how the bits are located in the data planes. Color models are described through their mathematical relationship to CIE XYZ (see section 1.1.2.1); the description is currently restricted to affine images of gamma-corrected color models (see section 1.3.3) which is sufficiently generic to represent all of the "physical" and "decorrelation" models used in multimedia (but cannot for example represent CIE L*a*b).

The code generation process starts out with an exact mathematical representation of the conversion to be performed, and successively transforms this representation towards executable code – by e.g. simplifying arithmetic terms, converting real arithmetic to fixed point arithmetic, replacing chains of "expensive" operations with look-up tables, and translating the bit packing required by the formats into shift and logical operations. The intermediate representation used for this optimization and code generation framework is somewhat unusual in that a unified directed, acyclic graph is used for all stages, and the graph is serialized into a sequence of instructions only after register allocation. (cf. [34]). Early experiments indicated that this representation is particularly suitable as it captures the "micro-parallelism" inherent to the problem very well.

# Appendix B

# Implementation notes: X Window System Extensions

## B.1 Real-time audio processing

The TIME extension allows clients to schedule execution of requests within the X server. One particular type of requests issued in this way are AUDIO requests to perform data transfer and transformation operations between different server-side sample buffers (see section 4.1.2.1). These are commonly used to allow mixing of multiple active audio streams into a single master playback buffer (see section 5.2). The latency introduced through software mixing is essentially determined by the worst-case execution latency of these mixing operations (see figure 5.2 on page 154).

To meet these tight timing requirements, scheduled requests are generally executed by a dedicated real-time thread within the X server. An accounting mechanism ensures that processing time is allotted equally to all clients and at the same time prevents the system from overload – usually it does not hurt if the true cost is grossly over-estimated (as CPUs are typically sufficiently fast to make computation cost for audio processing nearly negligible). However, the operations performed in the real-time thread on behalf of the client are quite simple (multiplication, accumulation, convolution); during experiments it was found that a trivial estimator that always assumes worst-case behavior (dataset too large to fit in L2 cache) does not over-estimate the common case (dataset already present in L1 cache) by a factor more than 2.

## B.2 Lock-free sample buffers

Sample buffers may optinally be placed in shared memory segments (see section 4.1.2.1). The mechanism allows X clients and servers operating on the same physical machine to exchange sample data without any X protocol interaction. Both sample values and the "base index" of the current window are placed in the same shared memory segment, thus the client can also perform operations such as shifting the window to a new position and inquiring the current position without generating X protocol requests. The semantics of the shared memory buffers is close to that of pure server-side buffers, but not completely identical. Both client and server must adhere to a strict protocol for access to the sample buffer to achieve desired behavior.

The memory locations holding the sample values are organized as a *ring buffer* of a fixed (power of two) size, and the size of the ring must be strictly larger than the logical size of the window represented by the sample buffer. Sample indices are mapped to ring indices in the usual way by taking their value modulo the ring size.

While the server implementation guarantees that shifting the window pointer to a new location is atomic with respect to concurrent access for all pure server-side sample buffers (which in particular means that any newly visible sample values are initialized to zero before they are accessible), it is for shared memory buffers the responsibility of the *writer* to provide this guarantee. Depending on whether the buffer is used for capture or playback, this role may be fulfilled by either client or server. The window may only be shifted forward while there can possibly be a concurrent reader, and the writer must adhere to the following access protocol[1]:

```
int new_base_index=atomic_load_relaxed(base_index);
while (shift_count--)
  ring_pointer[(window_size+new_base_index++) % ring_size]=0;
atomic_store_release(base_index, new_base_index);
```

while the reader must correspondingly execute:

```
sample_value_t samples[read_count];
int old_base_index=atomic_load_acquire(base_index), n;
for(n=0; n<read_count; n++)
  samples[n]=ring_pointer[(read_base_index+n) 5 ring_size];
int new_base_index=atomic_load_ordered(base_index);
```

The above access protocol introduces a benign data race in that the base index may be modified concurrently to the reader fetching sample data. In this case, some or all of the values fetched may be invalid, and the reader must retroactively discard them (i.e. set them to zero). To detect this, the reader must use the saved `old_base_index` and `new_base_index` values: They determine the base index of the window at some point before read access was started and at some point after read access was finished, respectively. From this, the reader can compute the intersection of the two windows and can therefore conclude which of the read sample values are *guaranteed* to be valid.

---

[1]The code example assumes that the variables `ring_pointer` and `base_index` point to the first element of the ring buffer and the storage location of base index of the sample buffer's window. The notation proposed in [7] is used for memory fences bound to shared variables.

# Bibliography

[1] Alsa project web page. http://www.alsa-project.org/, fetched on 2008-08-26.

[2] Bob Amstadt and Michael K. Johnson. Wine. *Linux J.*, 4, August 1994.

[3] Helge Bahmann. A Streaming Multimedia Extension for the X Window System. Diplomarbeit, TU Freiberg, 2002.

[4] Lorenzo Bettini, Sara Capecchi, and Betti Venneri. Double dispatch in c++. *Softw. Pract. Exper.*, 36(6):581–613, 2006.

[5] Jasmin Blanchette and Mark Summerfield. *C++ GUI programming with Qt 4*. Prentice Hall, 2006.

[6] Toni Bochmann and Frank Winkler. Ogg Vorbis Audio Komponente. Technical report, TU Freiberg, 2006. Internal report.

[7] Hans J. Boehm and Lawrence Crowl. C++ Atomic Types and Operations, 2007. ISO/IEC JTC1 SC22 WG21 N2145 draft paper for the C++0X standard.

[8] Olaf Borkner-Delcarlo. *GUI-Programmierung mit Qt*. Hanser, 2002.

[9] J.K. Bowmaker and H.J. Dartnall. Visual pigments of rods and cones in a human retina. *The Journal of Physiology*, 298:501–511, 1980.

[10] Cairo graphics web page. http://cairographics.org/, fetched on 2008-08-26.

[11] Carl D. Worth and Keith Packard. Xr: Cross-device Rendering for Vector Graphics. In *Proceedings of the Ottawa Linux Symposium*, 2003.

[12] Commission internationale de l'eclairage proceedings, 1932.

[13] Apple Computers. QuickTime File Format.

[14] W3 consortium. GIF specification. http://www.w3.org/Graphics/GIF/spec-gif87.txt, fetched 2008-08-24.

[15] W3 consortium. PNG specification. http://www.w3.org/TR/PNG/, fetched 2008-08-24.

[16] Eirik Eng. Qt gui toolkit: Porting graphics to multiple platforms using a gui toolkit. *Linux J.*, November 1996.

[17] Ralph Johnson Erich Gamma, Richard Helm and John Vlissides. *Design Patterns*. Addison-Wesley, 2007.

[18] Yôiti Suzuki et al. Precise and full-range determination of two-dimensional equal loudness contours. Technical report, Tohuku University, Japan, 2003.

[19] Flac format specification. http://flac.sourceforge.net/format.html, fetched 2007-12-20.

[20] Fletcher and Munson. Loudness, its definition, measurement and calculation. *J. Acoust. Soc*, pages 82–108, 1933.

[21] S.W. Golomb. Run-length encodings. *Trans Info Theory 12(3)*, page 399, 1966.

[22] Rafael C. Gonzalez and Richard E. Woods. *Digital Image Processing*. Prentice Hall, 2007.

[23] Gtk+ project web page. http://www.gtk.org/, fetched on 2008-08-26.

[24] Mike Spertus Hans J. Boehm and Clark Nelson. Minimal Support for Garbage Collection and Reachability-Based Leak Detection, 2007. ISO/IEC JTC1 SC22 WG21 N2670 draft paper for the C++0X standard.

[25] Matthew Hertz and Emery D. Berger. Quantifying the performance of garbage collection vs. explicit memory management. *SIGPLAN Not.*, 40(10):313–326, 2005.

[26] Holger Bönisch and Konrad Froitzheim. Server Side "Compresslets" for Internet Multimedia Streams. *icmcs*, 02:82, 1999.

[27] Daniel H. H. Ingalls. A simple technique for handling multiple polymorphism. In *OOPLSA '86: Conference proceedings on Object-oriented programming systems, languages and applications*, pages 347–349, New York, NY, USA, 1986. ACM.

[28] Coding of moving pictures and associated audio for digital storage media at up to about 1,5 Mbit/s – Part 3: Audio. ISO/IEC 11172-3:1993.

[29] Coding of audio-visual objects – Part 10: Advanced Video Coding. ISO/IEC 14496-10:2003.

# BIBLIOGRAPHY

[30] Coding of moving pictures and associated audio for digital storage media at up to about 1,5 Mbit/s – Part 2: Video. ISO/IEC 11172-2:1993.

[31] ITU. Digital compression of continuous-tone still images. http://www.w3.org/Graphics/JPEG/itu-t81.pdf, fetched 2008-08-24.

[32] Chad E. Fogg Joan L. Mitchell, William B. Pennebaker and Didier J. LeGall. *MPEG Video Compression Standard*. Kluwer Academic Publishers, 2000.

[33] Alan Bernard Bradley John P. Princen, A. W. Johnson. Subband/Transform Coding Using Filter Bank Designs Based on Time Domain Aliasing Cancellation. *Proc. of the ICASSP*, pages 2161–2164, 1987.

[34] Neil Johnson and Alan Mycroft. Combined Code Motion and Register Allocation Using the Value State Dependence Graph. In Görel Hedin, editor, *CC*, volume 2622 of *Lecture Notes in Computer Science*, pages 1–16. Springer, 2003.

[35] Kiia Kallio. Scanline Edge-flag Algorithm for Antialiasing . In Ik Soo Lim and David Duce, editors, *Theory and Practice of Computer Graphics*, pages 81–88, Bangor, United Kingdom, 2007. Eurographics Association.

[36] Ralf Keller, Wolfgang Effelsberg, and Bernd Lamparter. Xmovie: architecture and implementation of a distributed movie system. *ACM Trans. Inf. Syst.*, 13(4):471–499, 1995.

[37] Andrew Krause. *Foundations of GTK+ Development*. Apress, 2007.

[38] John R. Levine. *Linkers & Loaders*. Morgan Kaufmann, January 2000.

[39] Z.N. Li and M.S. Drew. *Fundamentals of multimedia*. Prentice Hall, 2004.

[40] Marco Lohse. *Network-Integrated Multimedia Middleware, Services, and Applications*. VDM Verlag, 2007.

[41] Marco Lohse and Philipp Slusallek. Middleware Support for Seamless Multimedia Home Entertainment for Mobile Users and Heterogeneous Environments. In *Proceedings of The 7th IASTED International Conference on Internet and Multimedia Systems and Applications (IMSA)*, pages 217–222. ACTA Press, 2003.

[42] Ralf Müller. Ein Audio Compositing Manager für das X Window System. Bakkalaureusarbeit, TU Freiberg, 2008.

[43] Peter Nilsson and David Reveman. Glitz: hardware accelerated image compositing using opengl. In *ATEC '04: Proceedings of the annual conference on USENIX Annual Technical Conference*, pages 28–28, Berkeley, CA, USA, 2004. USENIX Association.

[44] Bernd Oestereich. *Analyse und Design mit UML 2.1*. Oldenbourg Verlag München Wien, 2006.

[45] Keith Packard. A new rendering model for x. In *ATEC'00: Proceedings of the Annual Technical Conference on 2000 USENIX Annual Technical Conference*, pages 53–53, Berkeley, CA, USA, 2000. USENIX Association.

[46] Davis Pan. A Tutorial on MPEG/Audio Compression. *IEEE MultiMedia*, 02(2):60–74, 1995.

[47] Haoyu Peng, Hua Xiong, and Jiaoying Shi. Parallel-sg: research of parallel graphics rendering system on pc-cluster. In *VRCIA '06: Proceedings of the 2006 ACM international conference on Virtual reality continuum and its applications*, pages 27–33, New York, NY, USA, 2006. ACM.

[48] Havoc Pennington. *GTK+/Gnome Application Development*. New Riders Publishing, Thousand Oaks, CA, USA, 1999. Foreword By-Miguel de Icaza.

[49] Johannes Pfeiffer. Integration eines X11-GUI-Toolkits und eines verteilten Multimedia Systems. Bakkalaureusarbeit, TU Freiberg, 2006.

[50] Thomas Porter and Tom Duff. Compositing digital images. In *SIGGRAPH '84: Proceedings of the 11th annual conference on Computer graphics and interactive techniques*, pages 253–259, New York, NY, USA, 1984. ACM Press.

[51] Nico Pranke. Ein waveletbasierter Video-Codec – Algorithmen ud Implementierung. Diplomarbeit, TU Freiberg, 2008.

[52] Quicktime component creation guide. `http://developer.apple.com/documentation/QuickTime/RM/WritingQTComponents/MHCreating/MHCreating.pdf`, fetched on 2008-01-07.

[53] Iain E. G. Richardson. *H.264 and MPEG-4 Video Compression*. Wiley & Sons, 2003.

[54] Dale Rogerson. *Inside COM*. Microsoft Press, 1997.

[55] Robert W. Scheifler and James Gettys. *X Window System: Core and Extension Protocols: X Version 11, Releases 6 and 6.1*. Butterworth-Heinemann, February 1997.

[56] Ulrich Schmid and Klaus Schossmaier. Interval-based clock synchronization. *Real-Time Syst.*, 12(2):173–228, 1997.

[57] J. O. Smith. Physical modeling using digital waveguides. *Computer Music J.*, 16(4):74–91, 1992.

[58] Julius O. Smith. *Digital Audio Resampling Home Page*. http://www-ccrma.stanford.edu/~jos/resample/, January 28, 2002.

# BIBLIOGRAPHY

[59] Scott N. Steketee and Norman I. Badler. Parametric keyframe interpolation incorporating kinetic adjustment and phrasing control. In *SIGGRAPH '85: Proceedings of the 12th annual conference on Computer graphics and interactive techniques*, pages 255–262, New York, NY, USA, 1985. ACM.

[60] James CE Johnson Stephen D.Huston and Umar Syyid. *The ACE programmer's guide*. Addison-Wesley, 2004.

[61] Maureen C. Stone, William B. Cowan, and John C. Beatty. Color gamut mapping and the printing of digital color images. *ACM Trans. Graph.*, 7(4):249–292, 1988.

[62] Quentin Tarantino. Pulp fiction, 1994.

[63] Simon Thum. Ein nichtdestruktives Bildverarbeitungssystem. Diplomarbeit, FH Gießen-Friedberg, 2006.

[64] Tom Lane. Independent JPEG Group. http://www.ijg.org, fetched on 2008-08-26.

[65] Suramya Tomar. Converting video formats with ffmpeg. *Linux J.*, 2006(146):10, 2006.

[66] Jeff Tranter. Introduction to sound programming with alsa. *Linux J.*, 2004(126):4, 2004.

[67] Tiago Vignatti. Moving all the input code into a separate thread. Google Summer of Code 2008 project.

[68] Vorbis I specification. http://www.xiph.org/vorbis/doc/Vorbis_I_spec.html, fetched 2008-08-25.

[69] John Watkinson. *The MPEG Handbook*. Focal Press, 2001.

Die VDM Verlagsservicegesellschaft sucht für wissenschaftliche Verlage abgeschlossene und herausragende

## Dissertationen, Habilitationen, Diplomarbeiten, Master Theses, Magisterarbeiten usw.

### für die kostenlose Publikation als Fachbuch.

Sie verfügen über eine Arbeit, die hohen inhaltlichen und formalen Ansprüchen genügt, und haben Interesse an einer honorarvergüteten Publikation?

Dann senden Sie bitte erste Informationen über sich und Ihre Arbeit per Email an *info@vdm-vsg.de*.

**Sie erhalten kurzfristig unser Feedback!**

VDM Verlagsservicegesellschaft mbH
Dudweiler Landstr. 99          Telefon  +49 681 3720 174
D - 66123 Saarbrücken         Fax        +49 681 3720 1749
**www.vdm-vsg.de**

Die VDM Verlagsservicegesellschaft mbH vertritt

Printed by Books on Demand GmbH, Norderstedt / Germany